*The middle-weight champ in action...*

Rocking o[...] [...]ide to side, back [...] [...]ning speed, hea[...] [...] an endless suc[...] [...]s left snaking ou[...] rapidly than any striking rattler. His whole big, gray frame which had looked gross and clumsy became a smooth, elegant, fighting machine. Pit-a-pat, thuppety-thup, thud, thud, thud went the gloves; lead, counter, duck, jab, cross, uppercut.

The end came with startling suddenness. Some said it was a left, others a right; it was variously described as a hook, a cross, an uppercut, a straight trolley-wire right, but actually it was a whip punch that came from nowhere, half hook, half uppercut, a right that flashed faster than the eye could see, which lifted Billy Baker clean off his feet into a back somersault, landing him flat on his face.

# Matilda
## Paul Gallico

A BERKLEY BOOK
published by
BERKLEY PUBLISHING CORPORATION

# I

IN A DINGY office in a ratty building where in New York
Broadway and Seventh Avenue unite briefly for a block in
the middle forties, three men sat at three desks, strenuously
doing nothing except eyeing three silent telephones reposing
in front of them.

The place was so small there was barely space to pass
between the desks, the chairs and the filing cabinets. The wall
decorations, in addition to calendars furnished by the United
States Lines and Pan American Airways, identified the
professions of the three who had clubbed into the one-room,
hole-in-the-wall.

By far the most important position near the door was
occupied by a three-sheet poster of a prize fighter in ring
togs, in a fighting stance—left foot forward, gloved left hand
extended, right fist protecting the side of his jaw. Around his
middle was a broad gold belt intertwined with a sash in red,
white and blue. His face was amazingly unmarked by his
calling but revealed an expression of cold calculation. The
legend on the poster read: 'LEE DOCKERTY, MIDDLEWEIGHT
CHAMPION OF THE WORLD', and at the bottom in smaller type,
'Exclusive Management: P. Schwab, 1620 Seventh Avenue.'

Less conspicuous was a smaller poster in smudged black
and white showing a number of fighters of various weights in
aggressive poses, with nicknames involving either 'Kid',
'Young', or 'Kayo'. The text advised, 'Crowd Pleasers All'.
For Fighters that Fight Phone Bryant 3-6420. Patrick
Aloysius Ahearn, 1620 Seventh Avenue.'

Most modest was a bulletin board behind the desk in the

1

corner over which was lettered, 'Solomon Bimstein's Class Acts', and the slogan, 'BOOK WITH BIMMIE!' Pinned to it were a dozen or so signed photographs, all slightly faded, glossy prints of the kind used by night club and carnival acts for publicity. Amongst them were 'MIROSH AND MIROSHKA, MENTAL MARVELS'. Mirosh was presented in an ill-fitting dinner-jacket standing behind Miroshka seated blindfold in an evening gown five years out of date. 'VESUVIO THE HUMAN VOLCANO' appeared in East Indian costume and was seen expelling a blast of flame from his mouth. There were also photos of a sword-swallower, a blonde girl run slightly to flesh in tights with a python entwined about her, a humpsty-bumpsty knock-about act revealing three Italian acrobats fairly bursting out of their civilian clothes, a bedraggled-looking musical clown and a blues singer.

The three men at the desks in their pecking order were Pinky, Patrick Aloysius and Bimmie.

Pinky was shaped like a pear with his belt line well below his belly and never able to contain his shirt tail. He had a behind like a circus elephant and a large, jovial head with a jovial expression which masked considerable unjoviality. His eyes were small and greedy. His nickname had been derived simply from Broadway's adaptation of his first name—Pincus. Pincus Schwab, Manager of the Middle-weight Champion of the World, aristocrat of Room 309, 1620 Seventh Avenue.

Patrick Aloysius Ahearn as the veteran emeritus of the fraternity of Broadway fight managers was always respect-fully known by his first two names and was thus referred to by the sports columnists and boxing writers in their stories and in particular, with slight tongue-in-the-cheek, in the universally read column of Duke Parkhurst, 'The Duke's Deadline', in the morning *Mercury*.

Patrick Aloysius operated a profitable stable of mediocre club fighters who during their engagements specialized in getting themselves and their opponents hurt to the point where the audience, filing out of the arena, felt they had been suitably entertained. But Patrick Aloysius never let his kids take unnecessary punishment. With his silver hair, florid face whose sagging jowls were always lopsided due to the quid of tobacco occupying one cheek or the other, his unabashed vulgarity, experience, total cynicism and peculiar self-originated but workable set of ethics, made him something of a character on Broadway.

And the third and least was just Bimmie, on the fringes of show business, booking left-over acts from the bygone days of variety into cheap night clubs, carnivals or stags. But within his slight frame smouldered ambition. Someday Bimmie was going to be a big shot who would make Broadway sit up and take notice of him.

He was a pale, weedy youth of twenty-six with a prominent Adam's apple, bright intelligent blue eyes and wiry blond hair stacked in layers back from his forehead. When he became excited there seemed to be some electrical connection which made it rise. He would have been good-looking if his mother had not neglected to tape his ears to the side of his head when he was a child. His mouth was not without humor, his expression a combination of eagerness and avidity coupled with a desire to please. His speciality was gab. He was almost honest.

One of the telephones in the office exploded into life. Three hands twitched automatically but it was the one on Bimstein's desk. He plucked the receiver from its rest, cradled it between his left ear and collarbone while simultaneously picking up a pad and the stub of a pencil.

'Sol Bimstein's Theatrical Booking Agency—Book with Bimmie, speaking. Yeah, yeah, this is Bimmie, who's this speaking? Mr Matson? Matson's Carnivals? Where are you calling from, Mr Matson? Alabama? Yeah, yeah, Mr Matson, sure, Mr Matson, I'm writing it down, Mr Matson—a knife-throwing ack you want, Mr Matson? Sure, I got it. The greatest! You want a knife-throwing ack? Mr Matson, you never seen one like I got; it's like everybody's hair is standing up when they see it. You and I seen a lot of knife-throwing acks in our time, Mr Matson, but not never one where the feller is blindfolded, the girl spins on a wheel, the lights go out and he got flaming torches on the end of the knives, and the last one he throws at her nut got a firework inside which goes off, and when the lights come up she got an American flag over her head. Sensational isn't enough, Mr

Matson, it's stupendous. What? Yeah, yeah, sure I can get 'em for you. You're asking what are their names, Mr Matson? I'm just looking it up for you, I got it right here...'

The other two men were watching Bimmie fascinated, as he looked nothing up but instead doodled a large wheel on his pad with a girl spreadeagled on it, then scratched it out and rattled on—

'I got it right here, Mr Matson, but while I'm looking, maybe you'd rather have something even more sensational. It ain't only sensational, it's class. You know what I got for you, Mr Matson, just arrived from Austeralia? The world's greatest boxing kanagaroo, with The Bermondsey Kid that was once lightweight champeen of England...No, no Mr Matson, not the kanagaroo, The Bermondsey Kid, the trainer. He can show you the clippings. It's the greatest ack I seen since I been in the business. Because I wouldn't say that a knife-throwing ack was exactly in your line, Mr Matson, who is only booking strictly class acks and between you and me, Mr Matson, knife-throwing acks are a dime a dozen and went out with vaudeville and is not big enough for Matson's Monster Carnival that everybody knows from coast to coast, but a boxing kanagaroo ack is in, because you will not find any around, anyway, not like I got, because this is maybe the only boxing kanagaroo in the world who can go three fast rounds with the former lightweight champeen of England and then take on all-comers which is what gives the ack its kick, a great crowd-pleaser, Mr Matson, with a lot of comedy when the kanagaroo slugs some rube and knocks him ass over teakettle...What? I'm telling you, Mr Matson, this ain't no ordinary kanagaroo ack where they slap each other around and make like kidding with the gloves. This one's got a lef' like Cassius Clay and a right like Jim Frazeer. I was only talking to The Kid this morning and you should see the shiner he is wearing where his kanagaroo tagged him just working out yesterday...What's that, Mr Matson? I should stop calling it a kanagaroo? What is it *you* say? Oh, a kangaroo! Anything you say, Mr Matson. But I could fix you up with a six-sheet, the referee holding up the arm of the winner Matilda, the Fighting Kangaroo, and on the floor is a fighter he's just knocked cold with a straight right uppercut

4

to the jaw. Matilda's the greatest... What's that, Mr Matson? No, there ain't no dame in the ack. Matilda's the name of the kana... I mean, kangaroo, like you say. Not she, it's a he, Mr Matson, I can explain that later. But no kiddin', this Limey from England who trains him said he never seen a kanagaroo take to the game like this one, right from the time he started him as a kid, I mean when he was a baby. You can't go wrong with this ack, Mr Matson, them hayseeds will think he's a pushover on account of being a dumb animal, and will come up onto the platform when The Kid offers a money prize for anyone who can stay in the ring a couple of rounds with Matilda, so you've got plenty of comedy because The Kid says Matilda is so clever them rubes fall all over themselves trying to hit him and the audience falls right off their chairs laughing. What's that? Sure they could leave tomorrow, Mr Matson, if you will wire a couple of "C's" for transportation; yeah, yeah, just BIMBOOKING NEWYORK ... You will? I knew you'd recorgnize a class ack, Mr Matson, you won't regret this... Where do you want me to send them—Fair Grounds, Talawitchie, Alabama—Wait, I'm writing it down, Mr Matson—Talawhatsie? Oh, I see, Talawitchie... It's like they're there already, Mr Matson, and you better make it three "C's", on account maybe Matilda has to have a kind of special stall to travel on the train, being a... Sure, sure, Mr Matson, and thanks for letting me do you this favor...' He hung up.

Patrick Aloysius regarding Bimmie with intense amusement, fired a salvo of tobacco juice dead center into his gaboon and said, 'What the hell do you think you're doing? A guy calls up for a knife-throwing act and winds up with a boxing kangaroo?'

Bimmie was grinning all over his pale, thin face, his wavy hair standing up electric with excitement. He said, 'I ain't got no knife-throwing ack. But a boxing kanagaroo, I got. That was Matson the biggest carney show on the road. He's gone for three "C's", just for expenses. Boy, oh boy!'

Pinky said enviously and with some rancor, 'Looka, Bimmie, we give you orfice space here, you should book shows and acks. Fighters is our racket. You wanna book fighters, get yourself a licence and your own orfice.'

5

Bimmie ran his fingers through his hair and said, 'Are you kiddin'? A kanagaroo a fighter? It's strickly a ack. The Kid puts boxing gloves on its front feet and they mess about a bit, and I got me a booking. Wait 'til I tell Hannah!'

Pinky sneered, 'Three hundred bucks? Big deal!'

Patrick Aloysius was still shaking his head. 'A knife throwing act and the sucker winds up with a wild animal. I've heard everything now.' He began to lock his desk. It was almost six o'clock.

Pinky, too, was preparing to leave and took up the Russian style fur hat and the coat with the moth-eaten Astrakhan collar which made him look like a down-on-his-luck opera impresario, when the door opened and Lee Dockerty came drifting into the office.

Moody and more taciturn than usual, he was clad in tight pants and a leather jacket. Every line and bulge of his compact, muscular body looked like trouble. His hair was crewcut almost into a scalp lock, his expression as impassive and disinterested as that of an Indian. Indeed, he was supposed to have some Indian blood in him to which the sportswriters liked to refer when discussing his killer instinct, which actually was pure sadist, since when he got an opponent going, he would let him suffer a couple of rounds before finishing him off.

He merely nodded to Pinky without speaking, strolled past his desk to look at Bimmie's panel board, his hands jammed into the pockets of his leather coat. Having examined the photographs, he gave a snort of contempt which included Bimmie.

Pinky said, 'Listen, Lee, about this fifteen grand in Toledo. This Jimmy Cardo you should lick this time one hand behind your back.'

'Nah!' said Dockerty.

'What have you got to worry about a bum like Cardo for?' Pinky pleaded. 'You beat him twice and the ref that give a dezision against you in Kansas City that time was one of the boys. I'll get Michael Brian himself to go down with you and ref this one, and you'll be okay.'

'Nah!' said Dockerty.

Pinky tried again. 'Look, you don't got to make weight

6

for this one. You could take him in five rounds maybe, and be back in New York a couple a hours later.'

'Nah!' said Dockerty.

Patrick Aloysius, who had clapped a stained hat onto the back of his head and was working his cud of tobacco in his left cheek, said to no one in particular, 'Duke Parkhurst says Lee's afraid of Cardo since Cardo had him on the floor in Kansas City.'

Dockerty did not even bother to glance at Patrick Aloysius and merely annotated what one could do to Parkhurst.

Patrick Aloysius regarded Dockerty sardonically. 'You do it,' he said. 'I wouldn't want Parkhurst on my back.'

If he had not got under Dockerty's skin, he had managed to penetrate Pinky who said, 'Who the hell is Duke Parkhurst? A lousy sports columnist who don't know his ass from his elbow about boxing!'

Patrick Aloysius let them have one more thrust of the needle, 'That's right,' he agreed, 'who is this bum? Only the eyes and ears for a couple of million people.'

Dockerty produced another comment. Patrick Aloysius said amiably, 'That's what Parkhurst says you've got a rush of to the heart, when it comes to Jimmy Cardo.' He turned on Pinky, 'If he don't want to fight Cardo, why don't you match him with Bimmie's kanagaroo? Maybe that's more his speed.' He strode from the office leaving Dockerty disinterested and Pinky seething. The needle was Patrick Aloysius's specialty.

'Parkhurst! Parkhurst!' Pinky grumbled. 'We never had no break from that son-of-a-bitch and we don't need one.' He was referring to the fact that the influential sports columnist for the *Daily Mercury* had taken a dislike to Lee Dockerty for a number of reasons and had not hesitated to publicize them. One of these was that another newspaper had run a series on Dockerty's build-up fights that had led to his championship, alleging that most of them had been gangster backed phoneys. For reasons best known to himself Pinky had never brought suit over the articles and Parkhurst had been on him ever since.

Pinky tried again, 'Look, Lee,' he said, 'what's the matter

with you? That wasn't no knock-down in Kansas City, that was a slip. You were off balance. It was nothing but a lucky push. Cardo ain't going no place. He got stopped in his last fight. I got you a soft touch with him for fifteen thou' with your own ref. What do you say?'

Dockerty said it, 'Nah!'

Pinky said, 'For Chrissake, is that all you can do? Just stand there and say "Nah"?'

'Yeah,' said Dockerty, turned on his heel and walked out.

'That's what you get when you bring a son-of-a-bitch from nothing up to a champeenship. I made that boy,' Pinky grumbled. 'Suddenly fifteen thousand is peanuts. Anybody who wants to manage prize fighters should have his eyes examined.' He wrestled himself into his coat, one shoulder higher than the other, pulled his fur hat onto the back of his head and went out slamming the door, leaving Bimmie alone at his desk to meditate not so much upon the intransigence of Lee Dockerty but the marvel of being able to deal in such sums and the even greater magnificence of being able to turn them down. Dockerty had been paid a hundred thousand dollars as the challenger when he had won the title from Cyclone Roberts and two-hundred-and-fifty-thousand dollars in his first and only defense of it, when he had knocked out one Spider Ray in the third round—eight minutes work. And now fifteen grand was looked upon as small change. And Bimmie, who had not had a booking in two months, had been sitting there with ten dollars in his pocket to last him the rest of the week and nothing coming in, until with a stroke of genius he had managed to bamboozle the great Matson into taking Billy Baker, The Bermondsey Kid, ex-lightweight champion of England and Matilda his boxing kangaroo for three or four 'C's'—like Pinky had sneered, 'big deal'. But at least he would eat, and what was more, he would be able not only to ask Hannah out, but pick up the check as well.

# II

THE BREAK FOR Bimmie had actually begun the late afternoon of the day before, after closing hours, past six o'clock when he sat alone waiting for the silent telephone to ring and dreaming his dreams of grandeur to come when the name of Bimstein—for something or other, he was not quite sure what—would be on the lips of everybody on the Main Stem. And Hannah's father, Henry Lebensraum, would not be able to point a finger at his daughter and say, 'A bum like that you should want to bring into a good family? Maybe you have forgotten your mother was a Woolf, and from my side we had two rabbis, not to mention your great-grandfather who was killed in a pogrom in Cracow. A booking agent—phui! And when he takes you out, he borrows the money from you to pay for it.'

Sol Bimstein was the sixth in line of an octet of Bronx Bimsteins. His father, Leon, was a furrier's cutter, something of an aristocrat of the fur district off Eighth Avenue and the Thirties since good cutters of fur coats were almost priceless. His mother was a superior cook and famous the length of the Grand Concourse for her *blintzes* and *laftkes*.

The rest of the Bimstein brood, four boys and three girls, had ambitions towards respectable careers of one kind or another, all open to them through the marvellous free educational facilities provided them by the City of New York. But Sol just wanted to be a big shot and hob-nob with the great and the near great about whom he read with avidity in not only the daily papers, but *Variety* and *Billboard*, theatrical publications. His father wished to take him into the business but a block east of Eighth Avenue, Broadway glittered and Leon Bimstein was finally driven to the conclusion that Sol was nothing but a no-good bummer and a schmooser. 'Schmoos all the time. Talk, talk, talk, so maybe someday when you going into jail for a loafer, you can talk your way out of it.'

It was quite true that Sol, practically from infancy, could talk his way into or out of anything. In Public School 187 on Morris Avenue, he talked himself into jobs as monitor, distributor of books and supplies, which meant that he was never short of pencils, copy books, inks and erasers for himself. He did a minimum of work but talked his teachers into giving him passing grades. He talked his fellow students out of their pocket money and their possessions. In his spare time he haunted the bright light district between 40th and 50th Streets, talking his way into shows, prize fights and other entertainments. And when he emerged from P.S. 187, at the age of fifteen, he talked his parents out of making him go on to High School, invaded the famous theatrical building at 1600 Broadway and schmoosed himself into a job as office boy with Fink and Schulman, the biggest talent agents on the street.

Within two years he had argued himself into a desk of his own and the handling of half a dozen minor clients. Five years later, having learned all there was to know about the theatrical agency, talent and booking business, he bulled his way into office space with Patrick Aloysius and Pinky, stole as many of the clients of his employers as were greedy enough to be willing to listen to him and set up in business for himself.

There, however, he had stuck. For with the changing situation of the late sixties, even determination, burning ambition and the gift of gab were not enough. Business was rotten. Solomon Bimstein—'Book with Bimmie'—in four years on his own, had not succeeded in making much impression upon the Great White Way. He had remained not only a small fish but one gasping in a pond that was rapidly drying up.

The new tax law disallowing entertainment expenses to firms had been murder. Night clubs were closing down all over the country. 'Copacabana' was the only one left in New York running a show. Ed Sullivan had exposed most of the going acts twice over. People went to discotheques; nobody seemed to want live talent any more. Unless something came along, Bimmie was not going to be eating a great deal in the weeks to come. And even Hannah, his girl-friend, was getting fed up with paying for his meals when they went out together.

Bimmie did not at all see his predicament as just retribution or a punishment from on high for stealing a dozen or so of the clients of Fink and Schulman by promising them a bigger percentage and setting up in business for himself. Examples of any kind of retribution were rare on Broadway and besides, as Bimmie saw it, what he had done had not been crooked since during his tenure with Fink and Schulman he had served them well and faithfully, and been kept underpaid throughout on a salary which took him six months of saving to buy a suit of clothes, and from off the rack, too. Bimmie's yearnings encompassed custom-made shoes, monogramed shirts, tailored suits and overcoats and even special ties like the Main Stem big shots wore, bearing the coveted label of Sulka.

There were still plenty of ways of ringing up the Broadway jackpot: get yourself a stage star, pop singer or heavyweight champion to manage; back or produce a hit musical or television show, but for these—and this even the ambitious little agent had to admit—who needed Bimmie?

Alone there in the office he wondered even how much longer Hannah Lebensraum, whose father owned a prosperous men's suit and coat emporium and whose mother had been a Woolf, would continue to put any faith in his promises. Well, for one thing, at least she had kept him out of serious trouble. Bimmie was not above doing little business deals here and there which might have sounded just slightly off color if rehearsed by a lawyer for the complainant. But he had resisted the real temptations of the shadier side of the bright thoroughfare, the chances to make an easy buck the dirty way.

What had helped him to resist was the knowledge that as matters now stood between them, at least socially, he had a

11

long way to go before he could feel that he deserved such a girl as Hannah. He had no doubts that he would get there, that someday Mr Lebensraum would be proud to put his arm round Bimmie's shoulder and say, 'You know my son-in-law? The big . . .' Well, the big whatever it was that Bimmie was going to be. But for the moment he knew that he was nothing but a little *nebbich* of an unsuccessful booking agent whose prospects were growing dimmer by the hour.

He wished he could call Hannah and at least ask her out to a Chinese dinner. She had paid for the last two meals they had eaten together but even chop suey had suffered inflation and the kind of meal he would have to order for Hannah would come to at least ten dollars, which meant that he would be on short rations for the rest of the week. He had figured that twenty dollars to last him until a few pennies in commission might come in and he was too proud to ask help from his family.

In the lonely, silent quarter so far at the rear of the building that not even the roar of the traffic at the intersection of Seventh Avenue and Broadway penetrated wholly, he wondered how long he could continue to hold out.

He heard the distant clang of the steel door to the cage of the rickety elevator, far down the corridor and then footsteps approaching, curiously light ones, clicking down the wooden floor of the hall leading past the warren of offices on both sides. He lifted his head, hoping, but the footsteps passed his door and went on, then ceased seemingly hesitating, returned and stopped outside. The frosted glass panel showed a shadow, but a small one—no more than five feet two or three inches in height. Then slowly, as though activated by someone uncertain, timid or both, the door opened and a head poked through.

The head would have been astonishing and out of place in any setting except the one in which it now appeared. To begin with its ears had been crumpled until they resembled those of a fawn, the cartilage of the nose flattened and indented, almost like that of a pekinese. The lips, the upper one of which showed traces of some recent injury, when parted revealed gaps where at least two front teeth and several incisors were missing. The eyebrows were thickened and heavily scarred but they presided over merry, twinkling eyes—or rather one visible eye, since the other was almost shut by an enormous, no more than day-old, blue-black shiner. The head was as bald and the same general shape as the egg of an ostrich.

But there was nothing extraordinary about this visage to Bimmie. He saw them at least a dozen times a week, borne by indigent or punch-drunk ex-prize fighters dropping in to put the bite on Pinky or Patrick Aloysius for a couple of bucks. He was in no mood for charity at the moment, besides which it was not his department and before there could be any further incursion into the office he said, 'They ain't here. They've gone home.' He had been hoping against hope somehow that the shadow on the door might presage a customer of some sort.

The head, however, did not seem to be taken aback by this brush-off. Instead, the door opened completely, revealing a slender, compact, well-muscled little body. Its owner, from the wrinkles on his face, must have been in his late fifties. He was clad in a shabby suit and worn-out shoes. And thereupon Bimmie heard himself addressed in a strange voice with a strange lilt to it and in what amounted practically to a strange language. The words spoken were, ''Ave I the honor of h'addressing Mr Solomon Bimstein?'

Up to that time nobody had ever considered it an honor to address Bimmie and the compliment momentarily staggered him to the point where he forgot his natural caution and said, 'Yeah, yeah! Come in. I'm Bimmie.' But then the appearance of the man was such that his profession or ex-profession all came flooding back to him and he added quickly, 'But it ain't no use trying to put the arm on me.'

The little man stuck out a gnarled hand, most of the knuckles of which at one time or another had been broken, and said, 'Honored to be miking your acquyntance, sir, 'aving, so to speak, come on a matter of business. Allow me to hintroduce meself—Billy Baker, better known at 'ome and abroad as The Bermondsey Kid. Billy Baker from Bermondsey, you've 'eard of me no doubt.'

Bimmie had not and went back to the defensive

immediately. 'I don't handle fighters,' he said. 'Pinky—Mr Schwab and Paddy Ahearn ain't here. Come back tomorrow.'

'Cor!' said Baker. 'No need for you to worry yourself about that. I ain't pulled a glove on for twenty years, except with Matilda. Retired is what I am. Retired, undefeated lightweight champion of Great Britain. I've got me belt and I'm in the record books: 129 fights, 112 wins and only knocked out twice in me life.'

It all had a too familiar ring to Bimmie. It was going to add up to a touch. He said, 'Sure, sure! But I ain't got any time now, I was just closing down the office,' and he got up and began shuffling some papers about on his desk.

Billy Baker remained unfazed and did not budge. He grinned as ingratiatingly as his broken teeth would permit and said, 'But it ain't meself I've come 'ere to talk about, but me hact. You book hacts, don't you?'

Bimmie said, 'Your what?'

'Me hact . . . me and Matilda. We just signed off with the Robbins Brothers Circus, where we was the feature.'

'Oh,' said Bimmie and for a moment stopped rustling his papers. 'You mean you got a ack?' He was not much more enamoured of his visitor, even though the subject was not to be boxing. The world was full of acts; the problem was to find somewhere to place them. But as for the bite, there was still every chance that this would be forthcoming. Only a while ago he had read that the Robbins' Circus, one of the last of the smaller travelling shows in the U.S., had gone broke and had disbanded in Atlanta. To Bimmie this distant tragedy had only meant that there was now one less opening where he could place his clients.

But now his curiosity was aroused and he wondered what possible kind of act this beat-up little ex-prize fighter, and a Limey to boot, could have presented to have got a job, feature or not, with a show as good as the Robbins Brothers. The fact that it had fallen victim like so many others to rising costs and vanishing customers who were sitting at home watching circuses for free on their television sets, did not alter the fact that it had been a pretty good outfit.

And so against his better judgement he enquired, 'What kind of a ack? Who is Matilda?'

Billy Baker was now well inside the little office and said, 'Me boxing kangaroo. Billy Baker, The Bermondsey Kid, ex-Lightweight Champion of the British Isles and Matilda, his boxing kangaroo from Australia.'

For the second time Bimmie was startled and thrown by the wholly unexpected. 'A what?' he said. 'A kanagaroo? You mean an animal, like?' From the recesses of his past associations his rapidly revolving mind dredged up a faint memory that there at one time had been such a thing as an act involving a kangaroo. But they had gone out of fashion years ago and certainly such a one had never crossed the ken of his late employers, Fink and Schulman, so that Bimmie was unable to form any kind of a picture what it might be like. And, furthermore, he was not even too certain as to the nature of a kangaroo. But he thought it better to hedge swiftly and get out of whatever was coming, so he said, 'Oh, an animal ack...Nobody wants an animal ack today, on account you got to feed 'em and keep 'em. I had Rollo's Roller-Skating Bears, Captain Africa's Chimpanzees and Myer's Poodles. Nobody wants acks like that no more. There ain't no room in night clubs; vaudeville's been dead thirty years and you seen, like, how circuses are folding.'

Baker helped himself to a chair. 'Yes, Guv, I know,' he said. 'But Matilda's different. You ain't never seen nothing like him. 'E's not like any ordinary boxing kangaroo, not 'arf. 'E's as 'andy wiv his dukes as I was at me best, when I was champ.'

'Him?' Bimmie asked, suddenly intrigued. 'I thought you said her name was Matilda. Ain't she a lady?'

'Not 'im!' said Baker. ''E's no lidy. Anyway, lidy kangaroos don't box...it's just the males. Finest specimen you ever saw—five foot eleven when 'e stands erect and a wallop like a 'eavyweight.'

Bimmie was still confused. 'I don't get the Matilda part,' he said, 'if it ain't no lady.'

'Oh well,' Baker explained, 'on account of Australia where he come from. You know Australia? Grytest country

15

in the world! Got everything a man could want, except of course it ain't 'ome.'

'Oh,' said Bimmie, 'where's home?'

'Bermondsey, like I said,' repeated the little man, 'a bit of old London, down by the river.'

More confusion was growing in Bimmie's mind. 'Well, but what's that got to do with this Matilda not being no lady?'

'It's what I was just trying to tell yer,' Baker explained, 'because of 'is coming from Australia. You know, "Waltzing Matilda"—ain't you never 'eard the song? It's Austrylian slang from the bush. You know, carrying one's swag. See, it's the tucker bag the swagman carries along when 'e goes into the bush. It's called a Matilda. So one of these chaps had this little fella in 'is Matilda that 'e picked up somewhere outside Woolamaroo, where I reckon 'e'd got lost from 'is mum. 'E was just eight months old. So I bought 'im from 'im for a quid. I kind'a took to the little rascal and of course, I named 'im Matilda, see?'

Bimmie did not. He was hearing things that were so far beyond his ken that he could make no sense of them. However, Australia—which immediately became Austeralia—and kanagaroo, in his mind at least, seemed to hang together. He was fairly trapped now but also beginning to be intrigued.

'Look,' Billy Baker continued, 'that was years ago, and I was a lot younger then. I was with the Ralston and Fotheringay Circus, touring Down Under. I give an exhibition of rope skipping and light bag punching—I could play a tune on that bag—and I was a part of the clown act as well. I thought maybe when Matilda grew up I could make a boxing hact with 'im—'im and me.'

Bimmie was hooked now and there was nothing to do but go on, 'You mean teach him how to box?'

'Oh, you don't 'ave to do much teachin',' Baker said. 'Kangaroos are natural boxers. They get up on their 'indlegs and box with each other with their front feet, you know, out in the wild—in the outback.'

Bimmie did not know, but he was not going to let on so he merely said, 'Uh-uh.'

'It's a sort of a sexual thing with 'em, like,' Baker continued, 'for who's to be boss. If you ever seed a pair of 'em go at it together you could think you're sittin' in the bloomin' Albert 'All, back 'ome, watchin' the main event. So then all you got to do is to tie a pair of boxing gloves onto their front legs, keep your noggin out the way, and Bob's your uncle.'

'Bob's your. . . ?' Bimmie was finding himself more and more out of his depth.

'In a manner of speakin'. If you can get 'em used to you and learn to go along with their ways, you got yourself a hact. But right from the beginnin', after a year when I got Matilda used ter me and put the gloves on 'im, I could see 'e was somethin' different. 'E wasn't like no other kangaroo.'

Bimmie was now being swept along in the tide of narrative, his curiosity no longer containable. 'Whaddya mean, she—I mean, he—was different?'

Billy Baker's youthful blue eyes sparkled with remembrances, 'Why, it was just that this little fella was a natural. And what's more, 'e got interested in the gyme.'

'Interested. . . ?'

Billy Baker rose up from his chair to show what he meant. 'Like I said, boxing for a kangaroo is natural like, and maybe sexy. There's no science to it,' and here the little man windmilled both his arms. 'They just keep throwing lefts and rights until one of 'em lands a 'aymaker and the other one picks 'imself up and goes 'ome. Or they kicks with their 'indlegs, see?' Baker concluded, terminating his demonstration of throwing wild lefts and rights in the air. 'But not Matilda.'

'Not Matilda?' Bimmie echoed.

'Talent, yer see. 'E took to it for the fun of it. The first time I put the gloves on 'im, out in the barn in winter quarters, he didn't start waving of 'is arms like them others. 'E rocks back on that tile of 'is, sticks out 'is lef' and I walks into it. At first I thought it was a accident, so I come at 'im again and out goes that lef'. Pow! And after that, *ZAP*!'

'What was the zap?' Bimmie queried.

'The right. When I'd come around and got up off the floor, I knew I 'ad somethin' different. Matilda was more than just a boxing kangaroo. He was a natural of the manly

17

art. A bloomin' genius, that's what he was.'

'You mean that he could really box?' Bimmie queried in disbelief.

'Not yet, he couldn't. But he had the right idea first time out of the bag. 'E took a proper stance like it's in the book,' and here Baker demonstrated again, 'with his lef' out and his right cocked. Just to make sure I wasn't dreamin' I went round be'ind him and pushed his head down 'til his chin was tucked under his left shoulder. 'E never carried it any other way after that. He was like one of those kids that loves the game from the very beginning, for the game's sake. So I took 'im in and taught 'im all I knew.'

'Taught him what?' Bimmie asked. 'If like you say, he's a natural . . .'

'Oh, he was . . . he was that! But he was raw and a little crude, like all the kids when they start out, until they learn the finer points. He had a natural left jab but I taught 'im to hook and uppercut with it. He didn't know about jabbing in series until I showed him how you can stick a man silly just with your lef'. And there was a lot of work to do with 'is right, to keep it short. He had to learn how to slip a lead, duck, backpedal, tie a man up, feint with both 'ands, head and shoulders, block with his gloves, arms and elbows. Cor, he took to it like a duck tykes to another duck.'

Throughout this Billy Baker, now thoroughly warmed up, was illustrating every facet of the sweet science that he was telling about, leaping about the little office, bobbing, weaving, throwing half a dozen left jabs followed by a right uppercut, twitching his shoulder to indicate a feint, throwing up arms and elbows against imaginary punches, a little dynamo, a whirlwind of action, but one that hardly left him drawing a deep breath.

'I could see 'e was luvin' every minute of it. I was working meself up a act that was going to myke me fortune. I didn't even have much trouble with his hindlegs in the clinches.'

'His hindlegs?' said Bimmie. 'In the clinches? I don't get it.'

'Well, you see,' Baker explained, 'when a couple of 'roos get to fightin' and get serious about it, if one 'roo can lock his arms round the other one's neck and hang onto 'im, he tears

18

out his middle with his hindlegs, the claws of which are sharp as razors, like. He gets the power from that tail of his.'

'Jesus!' said Bimmie, the picture not being a pleasant one and his eyes rove to Baker's midriff which appeared quite intact. He said, 'How did you manage?'

Baker replied, 'I hardly had to. See, like I said, he wasn't like other 'roos. He liked to box clean, he did, according to the rules and regalations as I taught them to him. Besides which, he's got a luvin' nyture. Except when he's in the ring and hears that bell, he's got the disposition of a angel. When he knocks someone down and the fella gets up and clinches to hang on, Matilda wants to kiss. It's his good nyture assertin' itself, once he knows he's got the other fella going. He's got one or two cute tricks of his own he's worked up, when the referee ain't lookin'.'

'Hey, wait a minute!' Bimmie said. 'What referee? What fella? Do you mean to say your kanagaroo actually fights other people?'

'That's what I've been tryin' to tell yer,' replied Baker. 'That's why I got the greatest boxing kangaroo act in the business, because after I do me stuff with him; three fast two-minute rounds in which I do a lot of clowning and knock-about, the ringmaster he offers fifty nicker to anyone who'll step up, put on the mittens and stay three minutes with Matilda.'

'What if there aint no takers?'

'We got a couple of stooges who come up and know how to tyke falls. It's great for laughs.'

Bimmie said, 'Yeah, I guess so. But what if a legit wants to go for the fifty . . . What did you say they was?'

'Nicker . . . fifty quid. Pounds. We mike it more in America. He gets zapped, see?'

Bimmie gave a snort and for the first time was on more certain ground. He knew a lot about old-time carney routine. 'What do you do—move the sucker up against the backdrop and a guy slugs him with a tent stake from behind?'

Baker looked at Bimmie reproachfully, 'Mr Bimstein,' he said, 'there ain't no backdrop in a circus ring. Sometimes he zaps him quick; others he lets him stay for maybe a minute and a half for the fun of it and then he zaps 'im and that's it.'

'What if one of them should zap Matilda—a ringer, maybe?'

For the first time, Bimmie thought, the expression of extreme confidence on the face of the ex-boxer had altered and he seemed momentarily disconcerted. But he said, 'They don't. He knows too much for 'em, Matilda does. Maybe we could give you a little demonstration, so when you come to book our act...'

'Nix! Nix!' said Bimmie. 'I ain't said anything about booking your ack... I wouldn't know what to do or where to begin. There ain't no call for that kind of ack.' And to himself he thought: *Even if everything you've been telling me is true.* For he didn't believe a word of what the little man had been saying. 'How long you been in town, Mr Baker?'

'Two weeks, Mr Bimstein. I got Matilda in a stable down on 15th Street and Tenth Avenue, but I can't pay the rent. You want to know the truth? I ain't et for two days. I spent me last dollar on some lettuce and carrots and bananas for Matilda. He loves his bananas and 'Ershey Bars. They're his favorites. Mr Bimstein, did you ever see the look on the fyce of a animal that loves you and is hungry, and waits for you to feed him? It's enough to myke you want to curl up an' die. 'E don't know what's happenin' to him, Matilda, that he ain't got nothing to eat. All I'm askin' is that you help get us some work—maybe a smoker, or somethin'. Me and Matilda don't need much room.'

And the touch that Bimmie had been expecting all along never came. Billy Baker just sat down on the chair again and said no more.

'Where did you say you had him?' Bimmie asked.

'Ryan's stable, 781 Tenth Avenue. I'm sleepin' with 'im meself. You can find me there.'

Bimmie said, 'I can't promise anything. Business is bad. But I'll try my best.' And suddenly to his complete astonishment, he found himself reaching into his pocket and handing over ten dollars. 'Here,' he said, 'take this sawbuck. Get yourself a meal and the kanagaroo, too. I'll see what I can do.'

Billy Baker took the money with a nod of thanks and said,

'I'll pay it back when I can. You won't regret it. It's a great act.' Then he got up and went out, leaving Bimmie sitting at his desk, looking at the hand that had held his remaining ten-dollar bill. He was nearly broke himself and now he had given away half of his capital for the biggest snow job to which he had ever been subjected. He was hardly aware that what had really wheedled that ten-spot out of him was the expression on the ex-pug's face when he talked about the look in the eyes of hungry Matilda when no more bananas were forthcoming. Still, it was cheap at that, to get rid of him, for in his heart he likewise knew that there was not a chance anywhere around town of booking such a turn.

But that, of course, was before the telephone call from Mr Matson.

# III

RIDING DOWNTOWN IN the taxi two mornings later headed for Ryan's stable, Sol Bimstein was thinking that he had never made a better investment of ten dollars, forgetting the momentary charitable instinct that had prompted it. Never had bread cast upon the waters come floating back so rapidly or so richly, or so Bimmie would have thought had he stayed in school long enough to come in contact with the quotation.

Yet he was not entirely comfortable in his mind. For although he had booked Billy Baker's act to the famous Mr Matson who operated the last of the great travelling carney shows playing county fairs and had received the promptly wired expense money, the salary was still to be agreed upon. He was slightly worried as to just what it was that he had sold, and hence was on his way to have a look. If only half of what Baker had told him was true, Bimmie figured he could hit Matson for two hundred a week of which he felt entitled to keep half for being so smart. Things might be looking up for him, provided there was anything to the act.

The cab drew up at the corner of Tenth Avenue and 15th Street, where a sign over a building proclaimed, *'Ryan's Stables. Hay and Feed.'* Bimmie said to the driver, 'Maybe you'd better wait.' Now that he was realizing that he might have sold Mr Matson a pig in a poke, or rather a kanagaroo in a bag, for which he would never be forgiven, he was growing more and more nervous.

Mr Ryan was in his shirtsleeves encased in a miniscule cubicle of an office. The expression on his face was sour.

22

Bimmie asked, 'Where are they?... The fella with the kanagaroo?'

Mr Ryan poked a head attached to a scrawny neck out of the window and jerked it towards the rear of the stable. He said, 'They owe me forty-one dollars and fifty cents.'

Bimmie said, 'Sure, sure, I'll pay. I mean, after I've had a look at 'em.'

Mr Ryan jerked his head again and merely said, 'They don't go outer here without I get forty-one dollars and fifty cents.'

Bimmie made his way to the rear of the stable where there was a horse box. In the horse box Billy Baker was lying on his back in the hay, chewing on a straw and looking up at the ceiling. Crouched beside him was some kind of gray furry beast neither the size nor the shape of which Bimmie was able to ascertain at first as the light was dim.

As Bimmie peered around a corner of the stall, Billy Baker saw him and leaped to his feet, 'Oi, Mr Bimstein! Blimey, have you got a bookin' for us?'

'Maybe,' Bimmie replied cautiously, 'if the ack is like you say it is. I thought I'd better come down and have a look.'

Now that his eyes were becoming more accustomed to the gloom he was better able to distinguish the bundle of gray fur in the corner and his heart sank, for it did not look particularly impressive. On the contrary, Matilda hunkered down, doubled up and munching on a head of lettuce looked rather like an apologetic sheep. If that was all there was to a kanagaroo, Mr Matson was not going to like it.

But Billy Baker, unaware of the doubts engendered within his visitor, went over and gaily slapped the animal on the rump, 'Matilda! This is Mr Bimstein who's got us a job, come to see us. Get up and give 'im a kiss.'

A rustling and a stirring in the hay and Bimmie became witness to an almost terrifying transformation. Out of the cowering creature there arose a monster whose bulk seemed to fill the entire horse box. He stood, Bimmie figured, almost six feet in height, for he himself was five-foot-seven, and the beast towered both above him and his trainer. His pear-shaped body was balanced on two long hindlegs and tapering, curved tail as thick at the base as a hydrant. The

23

gray head set upon the sloping shoulders suddenly looked enormous, topped with what seemed to be donkey's ears. It had a long muzzle and soft brown eyes. Again Bimmie's lack of schooling denied him the comparison that Matilda looked exactly like Bottom wearing his ass's head in *Midsummer Night's Dream,* except that Matilda was more of a chimera than a dream.

And before Bimmie knew what was happening or could duck or flee, the creature came to him on one hop, put two powerful arms about his neck and deposited a long, wet kiss which covered practically all of the acreage of Bimmie's face. The animal smelt abominably.

'Stinks a bit,' said Billy Baker, as Bimmie recoiled. 'That ain't nothing to what 'e smells like when 'e gets excited and sweats. But you get used to it.'

Clasped in what seemed like an iron embrace, unable to escape the wet muzzle that was still implanting licks upon him, Bimmie panicked and cried, 'Ugh! Christ, get him off me!'

'There ain't nuffink to be afraid of,' Baker soothed, ' 'e wouldn't 'urt you. Not unless you was wearin' the mittens. 'E's got a 'eart full of love. Come over 'ere, Matilda, and give yer Daddy a cuddle.'

Obediently Matilda relinquished his hold upon Bimmie, took one hop and was embracing Baker. As a love-in it was a smashing success. Bimmie wiped his face dry and said, 'Jesus, he's big! What about his boxing?'

'Can't do it in 'ere,' Baker said. 'You want to 'ave a bit of room when he gets going. What's the gen?'

'If he's anything like you say he is,' Bimmie replied, 'I've got you booked into Matson's Carnival. They're down in some tank town in Alabama now, and I've got the money for expenses. But Jesus...!'

The expletive was only the surface indication of the thorough alarm that Bimmie was now experiencing. For the whole affair, the confines of the darkened horse box, Billy Baker was all but vanished within the embrace of the monstrous creature and the odour exuded by the amorous animal had taken on a nightmare quality and in a moment of terror the young booking agent only wished to extricate himself and flee. But then a combination of common and business sense took over. After all Baker and his beast had been a feature in Robbins' Circus—a check up in the back files of *Variety* had verified this. He had a sudden brainwave; a demonstration was called for. He remembered where Patrick Aloysius and Pinky sent their fighters to train and said to Baker, 'Listen, what about going up to Donohue's Gym and showing me the ack? I've got a cab waiting outside.'

'You're on!' Baker said. 'Wait 'til I gather me swag.'

For the first time Bimmie noticed that Matilda wore a leather nail-studded collar to which was attached a thin chain. Baker gathered a light and a heavy suitcase from the rear of the stall, took up the chain and said, 'Come on, Matilda!'

Bimmie asked, 'Will he go in a cab?'

'Sure,' Baker replied, 'he's used to it. 'E'll go anywhere as long as he's with me.'

They went out with Matilda progressing in short hops, propelled by that gigantic tail. Now, doubled over again, he looked much smaller and not quite so formidable.

Mr Ryan's arm was thrust out of the window of his cubicle. 'Forty-one dollars and fifty cents,' he said, and there was no nonsense in his voice. Bimmie was still so shocked from the menace of Matilda erect that he did not even argue. Besides which, sometimes one has to spend money to make money. He paid over fifty dollars from the money he had sequestered for himself from the expenses Matson had telegraphed, got the change and they went out to the waiting cab.

The driver eyed them dubiously. 'What the hell is that?' he said.

Bimmie said, 'It's a kanagaroo . . . a tame one. We want to go to Donohue's Gym—Broadway and 55th.'

'Not in my cab, buddy!' said the driver. 'Pay me what you owe me.'

'An extra finn,' Bimmie said. It was not the time to be penurious, besides which he was now devoured by curiosity to see Matilda in action.

The cab driver eyed Matilda once more. Down on all fours, he looked relatively harmless. 'Okay,' he said, 'but anything that thing does inside my cab, you clean up.'

'Anything he does inside your lousy cab could only do it good, like,' Bimmie said, 'with all them cigar and cigarette butts on the floor.'

'Nuffink to worry about,' Billy Baker comforted, 'he's as clean as you are. Holds his water like a camel. Never had a minute's trouble with 'im. In you go, Matilda.'

They all embarked. The cab driver shifted into gear and drove up Tenth Avenue shaking his head from side to side and muttering to himself.

Matilda crouched on the floor of the vehicle as docile as a lamb, a deeply introspective expression on his sheep's face.

'What's he thinking of?' Bimmie enquired.

'Search me. 'Ow would you ever work out what a 'roo was thinking of? A roo's a queer animal, stupid like, f'rinstance, when it comes to cars and they're sittin' in the middle of the road; let you drive right up their backs before they'll move. On the other 'and, you take when there's a dry spell and there ain't a drop of water in a river bed. Your old 'roo will know just where to dig to find that 'idden pool. Sometimes I fink Matilda's got everything figured out and others that 'e wouldn't know what time of day it was, if you asked 'im. But just you put the gloves on him and he knows what he's doin' all right.' Then he added, 'I'll bet I know what 'e'd like right now.'

Baker delved into the largest of the two suitcases which turned out to contain, amongst other things, a bag of carrots, a head of lettuce, a few Hershey Bars and a bunch of bananas. He broke off one of the latter, peeled it and said, 'Here you are, Matilda, have a banana.'

The introspective look turned to interest as though the trainer had read his mind. Matilda took the fruit with his front paws and munched it contentedly.

'Is that what you feed him?' Bimmie asked.

'Almost anything that ain't meat,' Baker explained, 'he's 'erbivorous. Any kind of fruit; bananas, apples, pears, carrots, lettuces. 'E'll eat grass or 'ay, likes bread and honey, and bread and jam. Chocolates, too—anything sweet. He gets a 'Ershey Bar after his work-out.'

26

Matilda had finished his banana and was making little happy clucking noises in his throat that sounded like, 'Uck-uck-uck-uck-uck-uck-uck-uck.'

'What's he saying?' Bimmie asked.

'That's his way of saying thank you,' Baker explained.

Bimmie ran his fingers through his wavy hair. He was still in somewhat of a daze; broke yesterday, today riding in a cab with a tame kangaroo and a booking in the offing. Well, that was Broadway.

Somehow Matilda seemed to have connected the unexpected gift of the banana with the presence of the young man in the cab, hoisted himself onto his hindlegs and climbed into his lap.

'Hey!' said Bimmie, half muffled in grayish-brown fur. 'What the hell is he doing?'

'He loves you,' said Baker. 'Didn't I tell you he had the disposition of an angel? You better let him sit there—he's sensitive.'

'Well then, for Chrissakes,' came in Bimmie's muffled voice, 'open the window!'

At 42nd Street they were halted by a red light. A man with a big cigar in his mouth, driving a luxurious Chrysler Imperial, drew up alongside, looked casually out of the window and caught sight of the tableau in the cab. He bulged his eyes, choked on his cigar, clapped his hand to his head and drove off through the red light.

They drew up at Broadway and 55th. The cab driver leaned back and said, 'Okay, here we are! Get that thing outer my cab, will yer? And don't forget that five bucks.' He sniffed suspiciously. 'What's that smell? He ain't done anything, has he?'

'Nah, relax!' said Bimmie. And then, 'Come on, Matilda, out!'

Matilda hopped docilely to the sidewalk. Bimmie paid off, leaving the banana skin as a souvenir to the already littered cab.

Baker handed Bimmie the chain and said, 'Here, you take him in, see? He's used to yer already.'

Their passage across the sidewalk created no stir since

27

New Yorkers would not consider anything less than a cerise elephant with wings on a leash worthy of a second look. But when they entered the old-fashioned brick building where whole generations of prize fighters had learned or practised their art, they caused a marked sensation.

# IV

DONOHUE'S FAMOUS GYMNASIUM was up one flight of rickety wooden stairs, presided over by Professor Donohue, an ex-light-heavyweight champion known as Gentleman Johnny from the days when boxers aspired to social standing as well as prominence in the record books. The Professor, now in his sixties, was a large, genial-appearing man with a flat, unmarked Irish face, since he had been clever in his day, hair parted in the middle and given to fancy waistcoats which he wore with an old-fashioned gold chain across the middle. He kept his roughneck clientele in awe and order by virtue of a quiet, unruffled elegance of manner.

For the most part, besides the fighters, the hangers-on and lay-abouts at Donohue's Gym were an unlovely crew: bald heads, bashed-in noses, bad teeth, cauliflower ears, thickened lips and eyebrow ridges, veterans of a thousand and one nights of glove fights. Here, too, were the cynical, twisted mouths, sharp narrow faces and gimlet eyes of managers, trainers, gamblers, hoodlums, boxing buffs, handlers, fixers, Broadway characters, Irish, Italian, Jews, Negroes, Krauts—all the mixture of races called Americans, the scum topping the surface of the fight racket.

But one and all, they knew the game, could spot a china chin or a slight touch of beagle in a fighter before he fairly had his hands up, catch that first slight give at the knees which indicates a man has been hurt, pick openings and spot weaknesses as they studied every move of the fighters sparring with their partners in the hopes of turning some minute discovery to financial advantage.

As Bimmie, followed by Billy Baker and Matilda, marched from the staircase which opened directly into the big room with its regulation-sized ring, there were some thirty or forty men and several women in the gymnasium, variously occupied. Trainers were bandaging the hands of boxers waiting their turn, massaging the leg or arm muscles of fighters who had come out or attending to an abrasion or cut with styptic pencil and dirty sponge. Others were merely standing about, their hands in their pockets, eyeing two heavyweights hugging in the center of the ring clubbing one another listlessly about the ears while the manager of one of them in a corner, his hands used as a speaking trumpet, was shouting, 'Alabanza, Jimmy, alabanza! Downstairs!'

Bimmie noticed that Patrick Aloysius Ahearn and some of his stable of fighters were there, too.

There was an immediate hubbub of astonishment at the trio and The Professor himself came strolling over to ascertain its cause. He looked upon Bimmie with distaste, since he did not like Jew boys, particularly young ones he did not know, and upon Matilda with repulsion. His voice was soft and gentle with an Irish lilt, but there was no mistaking his feelings when he asked, 'Might I enquire what the be-Jesus you think you've got there?'

His obvious hostility caused Bimmie to turn on the schmoos, 'It's a trained kanagaroo, Mr Donohue, he boxes... He's a boxing ack... I'm Sol Bimstein, you know, Bimmie? Book with Bimmie. I book acks... So this fella—meet Mr Billy Baker, ex-lightweight champion of England—has this kanagaroo and I got a chance to book him for Matson's Carnival—you know Mr Matson of Matson's Monster Carnival? So we just want a couple of minutes to try him out, so I can see how the ack works, so I can let Mr Matson know... It wouldn't take up any of your time, Mr Baker here's got all the gear... How about it, Mr Donohue?'

A look of pained disgust appeared upon the features of The Professor as he thumbed the armholes of his vest and replied, still softly, 'Well, now that you've asked me how about it, my reply would be—get that misbegotten creature out of here! He smells like a glue factory.'

The remarkable thing was that anyone would find

Matilda's odor noticeable over the general gymnasium stench of sweat-drenched woolen tights, soggy gloves, filthy canvas, rubbing alcohol, linament and garlic. Apparently, however, the mixture was agreeable to Matilda and stirred pleasurable memories, as did the sight of the ring, the sound of the scuffing of the big feet on the canvas and the thudding of the heavy gloves, for he stretched up on his hindlegs to his full height, his huge head sniffing the air appreciatively and a high gleam of interest moving into his eyes. This was more like it; maybe there was even going to be some zapping.

'Aw, now Professor Donohue,' Bimmie argued, 'give him a chance! He ain't no ordinary kanagaroo, he's a fighter, like. He boxes just like anybody else, only better. He's a genus at it. This is a gymnasium for boxing, ain't it? I'm willing to pay. See, I got to book this ack and...' He was stuck here momentarily because he did not want to hurt Billy Baker's feelings by saying right out loud in front of him that what he really wanted to find out was whether Baker had been telling the truth, or the whole thing was nothing but a con. But he picked up speed again immediately with, 'See, if this kanagaroo is as good as Mr Baker says he is, I can hit up Mr Matson maybe for a couple of C's more for the ack ...Maybe you'd be enjoying yourself, too, Professor, because where will you see an ack like this for free?'

Professor Donohue turned away from Matilda and Bimmie and remarked over his shoulder, 'In case you haven't got that thing out of here within thirty seconds, you will be finding your ass on the sidewalk.'

Intercession came from a wholly unexpected source. The sardonic Patrick Aloysius Ahearn had been standing over on one side, chewing on his cud, an amused expression curling around his chops. He addressed The Professor out of the side of his mouth, 'Why don't you let him go a couple of rounds, Johnny? If that thing can fight it could be a change from a lot of these bums around here. I'd like to see it meself.'

The Professor scratched his head. An appeal from a fellow Irishman and an old pal was something quite different from an invasion by a fresh young wavy-haired Jew boy accompanied by a broken-down pug and an animal that looked like a donkey that had been put together back to front, barging in without so much as a by-your-leave. He said, 'Do you think so now, Patrick Aloysius?' falling into the Irish way of speech. 'Well, now, maybe there could be something to what you say.'

The veteran fight manager fired his accurate jet into the gaboon just beneath the ring posts and said, 'Bimmie's okay. He works out of our office. He books acts for show business. Give him a break.'

Donohue stood for a moment, his thumbs still hooked in the armholes of his fancy vest, his head cocked on one side, studying the pair who were still locked in an embrace in the ring. Finally he said, 'Come on you two consenting males! Blow, before you're charged with public indecency.'

The heavyweights untwined and looked down at The Professor with uncomprehending bovine eyes. One of them muttered gutterally, 'Huh? What's that?'

'Get lost!' ordered The Professor. 'Move the love-in. We need the ring.' He turned and nodded to Bimmie, 'Okay. Go ahead.'

Billy Baker said, 'You want to see the whole act?'

Bimmie was still unnerved and his reception at the gymnasium had not helped. He replied, 'Nah. You said he could box. Let's see him do it.'

'Okay,' Baker agreed, 'we'll go three minutes, like, without all the falls and tumbles I do. We can play it straight or clown it up.'

'Just get going,' Bimmie said, 'before we get trun out of here.'

Baker took off his coat, rolled up his shirt sleeves and once more delved into the large suitcase from which he removed two pairs of well-worn sparring gloves.

Peering over his shoulder, Bimmie saw further paraphernalia—a red fright wig, a clown costume, an Australian flag. Bimmie was not the only one peering. Showing every indication of mounting excitement, Matilda came over sniffing, 'uck-ucking' in his throat as he nuzzled gently amongst the contents.

Baker said, 'He's gettin' worked up. He loves a good go. He always performs better before a new audience. Here, Matilda, you'd better have another banana.'

He peeled and handed over the fruit saying, 'It keeps him from gettin' too excited.' While Matilda chomped, Baker deftly laced one pair of the gloves onto his forefeet then pulled the other two onto his own hands, which one of the handlers in the crowd automatically did up for him.

When they were ready, Baker clapped his leathered hands together and said, 'Okay, Matilda, in you go!'

The kangaroo did it on the fly, that is to say in one hop from the ground clean over the top rope—a leap of some fourteen feet. The crowd around the ring gasped and a number of watchers backed away slightly. Someone said, 'Jesus Christ!'

Matilda did not sit down, but stood in the opposite corner erect, with both arms along the ropes, his massive chest thrust forward. His nostrils were twitching, so were his ears, as his head turning from side to side searched for an opponent. Inhuman, he looked in some ways grotesquely human and horrifyingly capable.

Baker went around the side of the ring and climbed up the steps and through the ropes in the facing corner. He called out, 'Anybody care to referee? 'E's been taught to break nice and clean.'

Nobody moved. Patrick Aloysius grinned and said to The Professor, 'Go ahead, Johnny, you're the best referee in town.'

The Professor said succinctly, 'Not for all the hop in China! Get in there yourself, if you want to.'

'Never mind,' said Baker, 'we don't need one. 'Im and me know each other. Somebody hold the watch. Three minutes. Okay, let her go!'

The bell clanged. Baker was up off his stool in classic, pro's boxing pose, shuffling forward. It took Matilda just one abbreviated bound to reach mid-ring. They actually touched one another's gloves in the time-honored gesture and thereupon followed one of the prettiest little exhibitions of what has been called the Manly Art, that even the oldest old-timer could remember.

Matilda boxed indeed as Baker had described to Bimmie, his left out, his right cocked, his long chin tucked under his left shoulder. But that was only the beginning of his science

33

as he displayed it. Rocking on his tail, he could move from side to side, backwards or forwards with lightning speed, head and body were never still in an endless succession of shifts and feints, his left snaking out more rapidly than any striking rattler. His whole big, gray frame which had looked gross and clumsy became a smooth, elegant, fighting machine. Pit-a-pat, thuppety-thup, thud, thud, thud went the gloves; lead, counter duck, jab, cross, uppercut with Baker weaving in and out equally adept at avoiding blows, slipping or blocking them.

Ordinarily before a minute was up he would already have taken two or three comedy falls, but upon this occasion he was intent upon showing off Matilda's paces rather than his own before such an audience of experts. He contented himself with keeping out of trouble.

Matilda weighed in the neighborhood of a hundred and sixty pounds to Baker's a hundred and forty and moved his bulk around like a bantam, picking off punches in mid-air, tantalizing Baker just out of range, then gliding in with a series of jabs that would have torn the trainer's head off if he had not been equally quick to slip in underneath and tie up Matilda for a moment.

He said, 'Break!' as the referee would have said it and indeed Matilda broke clean, stepped back and then advanced again. The sheeplike expression about his muzzle had changed to a mixture of keen concentration, interest and pleasure. It was obvious to the astonished and murmuring crowd around the ring that not only had he been beautifully trained, he was also enjoying every minute of it.

The end came with such startling suddenness that when they were all discussing it afterwards no one really was able to describe quite accurately what had happened. But during one of the exchanges, a voice from the crowd cried out, 'Come on, Billy lad, let's see you 'it 'im a good one for Bermondsey!' Obviously some Limey old-timer who recognized The Bermondsey Kid was in the group of spectators.

To hear his own language unexpectedly, distracted Billy Baker and for a fraction of a second he lowered his guard and turned his head. That was all Matilda was waiting for.

Some said it was a left, others a right; it was variously described as a hook, a cross, an uppercut, a straight trolleywire right, but actually it was a whip punch that came from nowhere, half hook, half uppercut, a right that flashed faster than the eye could see which lifted Billy Baker clean off his feet into a back somersault, landing him flat on his face.

Fortunately the upward force of the blow was robbed of some of its sting by the fact that it first came in contact with Baker's chest before it landed underneath his chin, or it might have torn off his head.

Bimmie emitted a howl of laughter, 'Boy, what a gag!' he said. 'That's the funniest gag I ever seen!'

But nobody else was laughing and Patrick Aloysius Ahearn was heard to murmur, 'Holy Mother of God!'

The Professor removed his thumbs from his vest to his hips, walked over to the ring and inspected the prone Billy Baker carefully. 'Brother!' he said.

Matilda had retired to his corner and waited, propped up on his tail. Somebody amongst the ringside onlookers was counting, '. . . Fifteen . . . Sixteen . . . Seventeen . . .'

Billy Baker stirred, quivered, got his knees under him, struggled to get up and finally managed, but he was still out, his legs shaking. As Matilda emerged on a hop from his corner, Baker could only grope for him half blind, in an effort to grab him and hang on.

Bimmie was still slapping his sides with glee and amusement. 'Boy,' he said, 'Mr Matson's gonna love this! It's the greatest ack I ever saw. It's worth three hundred a week.'

But the 'ack' was not over yet and now took a wholly different turn. For it suddenly became quite soppy as Matilda pulled his trainer into his powerful arms and began to kiss him with loving and affectionate tenderness.

The big, wet muzzle deposited big, moist kisses upon the face of the dazed Baker. They began to revive him and in fact he was able to shake his head to clear it and then push himself out of the clinch. At this point the fellow holding the stopwatch clanged the bell to end the round. It came just in time for Matilda, having noticed that his victim was returning to his senses, was preparing to zap him for keeps.

But the sound of the gong turned Matilda back to his

corner, where this time he sat down on the stool, with his arms stretched out along the second strand of rope.

Somehow Baker managed to stagger down the ring steps, over to Bimmie. He thrust out his arms and said, 'Get these bleedin' gloves off me 'ands!'

Bimmie unlaced them and said, 'Boy, what a gag! You were great! This ack will knock 'em in the aisles!'

But Baker hardly heard him for he was not entirely with it yet. He managed to mumble, 'Has anybody got any smellin' salts?'

One of the professional seconds nearby produced a small bottle of ammonia crystals from a pocket, unstoppered it and passed it under Baker's nose several times until the glaze went out of his eyes and he shook his head and said, 'If you don't watch that barstid every second, he'll bloody well kill you. I suppose he's got to have his reward now, for ruddy near tearin' me head off. Here you, Matilda! Come down out of it!'

He reached into his bag and brought forth a Hershey Bar, unwrapping it as Matilda, with one bound cleared the ring onto the floor, accepted the candy with every appearance of delight and once more collapsing into the insignificance of all four feet to the ground, quietly began to devour it.

The ringside was alive with the mumblings, murmurings and mutterings of the spectators:

'What did he hit him with?'

'Boy, what a punch!'

'Did you see them feints?'

'That right didn't move no more than six inches.'

'It was a left!'

'I wouldn't send one of my boys in with that, he knows too much.'

The voice of the Englishman was heard saying, 'Did you see? Baker never laid a glove on him!'

Another voice added, 'It ain't human!'

Baker had packed away his paraphernalia in the suitcase and had Matilda by his lead chain as The Professor came over remarking, 'That thing's dangerous. There'll be no charge, but I'd appreciate your getting him the hell out of here before somebody gets hurt.'

Bimmie, who was the only one who had not an inkling as to what really had happened, was still in the euphoric state engendered by what he thought had been Baker's acrobatic flip-flap. He said, 'Yes, Professor Donohue. Like you say, Professor Donohue. Thank you, Professor Donohue. Oh boy, what an ack!'

On the way out Patrick Aloysius Ahearn stopped him. He said to Bimmie, 'You wouldn't care to sell a piece of him, would you?'

Bimmie replied, 'Are you kiddin'? It's my ack. I discovered him. I'm signing him up. An ack like that is worth four hundred a week.'

Patrick Aloysius said, 'Maybe. But if I had a middle-weight who could punch like that...'

The same cab was still drawn up at the curb, not yet having picked up another fare, for it was a slack time of day. When the driver saw them emerge, he reached back, opened the door of his vehicle resignedly and said, 'Okay, okay, it might as well be me again. And don't forget the extra finn. Where do you want to go this time?'

They all piled in and Bimmie gave his directions, 'The Pennsylvania railroad freight and shipping yards at Eighth Avenue and 25th Street.' To Baker he said, 'Congratulations! You got a job. You're on your way to Talawitchie, Alabama.'

It just so happened that nobody of importance had been training at Donohue's that morning. None of the sportswriters or anyone from the Press had been present, and so the following day there was not a line in any of the papers regarding the incident of the boxing kangaroo and what eventually the hangers-on and wise-guys around the ring concluded was probably an accidental swipe of its paw, and not the fastest and most cannily delivered right-hand knockout they had ever witnessed.

That night, with Billy Baker and Matilda together in a special freight car waybilled and *en route* to Matson's Monster Carnival, Talawitchie, Alabama, Bimmie was regaling Hannah Lebensraum at Gallagher's Steakhouse on

52nd Street off Broadway, with his triumph. They were having broiled lobster and beer—a tremendous extravagance.

Bimmie finished his narration of the events of the last two days with, 'So I got Matson on the blower again, told him the ack is on the way and conned him out of another hundred. Four hundred a week; two hundred for them and two hundred for me. Baby, just you watch Bimmie! Wait 'til you tell your old man about this.'

But Hannah said, 'Is it honest, Bimmie? I thought agents were supposed to take only ten per cent.'

Her query came as no surprise to Bimmie. Of course, Bimmie considered himself on the level and would not deliberately set out to swindle anyone, except in the booking business sometimes one found oneself in a situation where it was better to keep one's mouth shut than to complicate matters by producing all the facts.

As an uptowner, Hannah, a graduate of Hunter College and holding down a responsible secretarial job, had the uptowner's suspicion of what went on in that area between 39th and 52nd Streets whose incandescent glow nightly lit up the mid-town sky.

They had met two years ago at the Bar Mitzvah of Bimmie's younger brother, had fallen in love and, to the dismay of Hannah's parents, had been going together steadily ever since, or as steadily as a boy whose income fluctuated downwards more than upwards could manage.

What Hannah saw in Bimmie was a mystery to her elders who were not with it when it came to the dreams and aspirations of the younger generation. Hannah had been touched by the depth of Bimmie's eagerness to succeed. His faith in himself and that she would someday share in some manner all the fabulous wealth, success and popularity he would achieve had inspired a like faith in her. In short, Bimmie had something—if nothing else, courage which her father preferred to refer to by the denigrating Yiddish word *chutzpa,* standing for Jewish nerve. Her only worry was that on the road to success he might somewhere be led off the straight and narrow.

For Bimmie, Hannah was the enchantress born and

38

destined to wear his jewels, his mink coat, ride in the back of one of his several Cadillacs and preside over his Riverside Drive duplex when that day should arrive. She was all the Old Testament famous—Sarah, Ruth, Rebecca, Queen Esther—rolled into one. In spite of being city born and bred, she had the freshness of an Israeli sabra; thick, dark hair framing an oval face from which shone the haunting eyes of her race. Bimmie yearned to dress her hair with diamonds and entwine her throat with pearls. However, he had to defend himself.

'Ten percent?' Bimmie repeated. 'Look, Hannah, so where would they be today if it hadn't been for me? Still down in that lousy stable chewing straw, or maybe starving to death. Two hundred per is as much as he was getting from Robbins' Circus, so what's he losing? Okay, so an agent gets ten percent, but a fighter gets cut up the middle. Matilda's a fighter, ain't he? So where's the difference? What I ought to have is sixty percent, but I'm only taking fifty, ain't I? I'm doing him a favor.'

This was not exactly true, but so close as to make no never mind. For one lousy percent Bimmie had not felt it worth while to enlarge on it. Before he had sent Baker and Company off to the unpronounceable place in Alabama, he had first written and signed a contract with him in which he, Bimmie, was down for fifty-one percent and Baker for forty-nine, explaining it thus: 'See, it's really fifty-fifty. You get the extra one percent in cash. It's only so when I make a deal for you I can talk like from fifty-one percent, now.' The Bermondsey Kid had been both too befuddled and too grateful to argue, or realize that he was signing away control of the act.

'And anyway,' Bimmie continued, 'didn't I split the last cent I had with him when I was broke and didn't know where my next buck was coming from? Ten bucks I hand some guy I'm never gonna see again, after he tells a tale about having a kanagaroo that's starving.'

Hannah turned her dark, exquisite eyes upon her suitor. His arithmetic had managed to confuse her and maybe Broadway business was different, besides which for the first time in six months he was paying for a meal for the two of them, and what a meal! And she knew it was true that Bimmie had parted with half his capital out of sheer charity.

She placed her hand upon his across the table and said, 'That's right. So you did. You're a good boy, Bimmie, at heart.' Nevertheless, she was still slightly worried.

# V

Nature marks its four seasons with the buds of spring, summer foliage, autumn's palette and winter's icy winds and snow. For the sportswriter the changes are distinguished rather by the sounds connected with the games as they vary from the crack of bat upon ball, the thunk of the forehand drive or overhead smash, the click of a golf ball taking flight, the rallying college cheers from the grandstand as the football behemoths grind one another's faces into the turf, and the ringing of the steel blades of hockey skates from the ice.

The indoor track season, which produces the climax of the winter solstice, had come to an end. The baseball clubs had taken off on their annual migration to the south for their spring training and with the hegira Duke Parkhurst, sports editor and columnist of the *Daily* and *Sunday Mercury,* treated himself to half holiday, half work by boarding a jet for San Antonio, Texas, to inspect The Mets Baseball Club at their practice.

What might have been a less than innocuous ending to a voyage that terminated not in San Antonio but near the small town of Camauga, Mississippi, was due to the good fortune that Reckmeyer Field, a large Army Airforce training center, was located just twelve miles south of Camauga. Hence when two engines of the big jet unaccountably began overheating and had to be quenched at eight o'clock at night, some fifty miles north of Camauga, the pilot dabbed the beads of sweat from his forehead, said 'Thank you God!' and radioed the Airforce for permission to land.

He did so without so much as a jar to disturb his passengers and after communicating with his base in New York, informed them that they would have to stay there until the following morning when another ship would be flown down to take them on. The Army offered the hospitality of its Officers' Club and an unoccupied barracks and thereafter the travellers were left to amuse themselves as best they could.

Mr Parkhurst was not pleased. Captive to his fellow passengers, most of whom were business bores, as well as young Airforce trainees, he saw himself in for one of those evenings of interminable questions about this or that ball player, prize fighter, tennis champion, golf aspirant, not to mention equally endless analyses of the pennant chances of the various ball clubs. All of these things Parkhurst produced in his daily columns. It tired him to tears to have to talk about them as well. He was not to be found in hotel lobbies with other sportswriters catching phantom forward passes, demonstrating the overlapping grip or throwing up statistics.

While the others were settling themselves and still discussing the excitement and fortuitous nature of the forced landing, Parkhurst was looking for a place to hide. The camp movie advertised for the evening was no help. It was a picture he had seen and under no circumstances could be lured into sitting through again. But as he wandered about cursing the situation in which he found himself—he would have to wire his office to use one of the emergency columns he always left behind for the next day—he came upon a large red and yellow poster affixed to one of the recreation buildings. Six-inch high lettering invited, 'COME TO THE MATAPOSA COUNTY FAIR!' and in smaller lettering, 'County Fair Grounds, Camauga. April 1st to 15th.' The poster further detailed the attractions and exhibitions to be encountered: races, fireworks, livestock judging, hog calling and pie baking contests and in again more prominent type, 'MIDWAY'.

The name of the town, Camauga, rang a bell in Parkhurst but it was so faint he could not remember where he had heard it before. He was certain that he had never previously been to what was probably a rural county seat of no more than six or seven thousand population. But the whole idea of a county fair, and particularly the word '*Midway*', attracted him strongly.

In his youth Parkhurst had been an up-state New York boy and no stranger to these county jamborees, with the result that for all of his irreconcilable distaste for baddies he had a soft spot for the carney people, the concession operators, barkers, shills, pitchmen, grifters and cappers who specialized in separating the suckers from their nickels and dimes. There was a certain naïve innocence and blandness combined with shameless effrontery in the palpably slanted games that the players could not possibly win that engendered a feeling almost of affection in Parkhurst. They worked so hard to get the rubes to 'Step up and try it!—Everyone a winner!' that it almost amounted to a delicious kind of art, particularly since it was aimed at boobs who thought they could get away with a forty-dollar radio, camera, clock or other valuable article for a ten cent investment.

As he read the poster, in his ears were already ringing the cries of the outside talker, 'Hurry, hurry, hurry! Come right up! Bring your girl! The show's about to begin! See Geeko the wild dancing bushman and the tatooed man!' He imagined the grind organ of the merry-go-round and the snapping and popping of the .22 rifles in the shooting booths. He felt a great longing to go and get taken himself. Before very long he encountered a young pilot lieutenant from New York, goggle-eyed over meeting him, and had little difficulty in persuading him to lend him his car. If it was like any county fair he had ever known, there would be plenty to eat in the form of hot-dogs, hamburgers and candy floss, and so not even accepting the invitation to dine at the Airforce mess, he folded up his great bulk into the lieutenant's machine and drove off in the direction of the Camauga fair grounds.

Duke Parkhurst was an important and powerful man in New York. Everyone from local millionaires and socialites to the Mayor and even the Governor were pleased to know him and ready to do him favors. True, he had free tickets to give away for various sports events, but this largesse could not account for his popularity, nor was the circulation of the *Mercury*—close to two million copies—the reason for the tremendous amount of weight he pulled in the world's toughest and most blasé city. There was no more than a handful of genuine celebrities who could impress New York. Duke Parkhurst was one of the most impressive.

To begin with he gave the physical impression of power from the space he took up. For he was a huge man, six foot four-and-a-half inches in height, with shoulders like barn doors, mounting a massive head. He had whimsical eyebrows set in a broad face and he sported a moustache like an old-fashioned Irish precinct captain. He had hands like hams and a pair of penetrating eyes that could bore twin holes through a phoney at a distance of a hundred yards. He was big, big, big and, what is more, almost universally liked, except by his enemies and they respected him.

It was probably his uncompromising and aggressive honesty that was the well-spring of his influence. He was constantly at war with the crooks, fakers and racketeers and lately had even taken on the Mafia which had exhibited signs recently of expanding its business interests to take in professional games.

His daily column, 'The Duke's Deadline', a thousand words a day, ran down the right side of the lead sports page. Often hands turning to it were so nervous that the paper rattled as they fumbled their way to that section to see what happened, was happening, or was about to happen in that world. Three lines of praise were considered better than a certified check; a hint that all might not be on the up-and-up in a forthcoming contest would send odds tumbling and bookmakers scurrying.

Hackies, elevator boys, cops, doormen, head waiters, shoe clerks, short-order cooks and all-night disc jockeys swore by him and the children of the poor adored him. For he organized and directed the *Daily* and *Sunday Mercury's* 'Free Food Fund for Hungry Children', collecting hundreds of thousands of dollars annually, every penny of which was sent into a fund to alleviate malnutrition amongst the young in the slums of the city.

Parkhurst was also the originator and promoter of the

biggest amateur boxing tournament in the metropolitan area, the Diamond Belt challenge, in which for a gilded belt with a diamond chip in it some four thousand young amateur hopefuls punched, slugged and bashed one another for a month of eliminations before meeting for the finals in New York's Metropolitan Arena before some twenty-five thousand, wildly enthusiastic fans. The Diamond Belt champions were the élite of the amateur boxing world and usually turned professional. The success of the tournament and the final sell-out night in the Arena was a signal to advertisers that if they wanted to put over a message to men, the columns of the *Mercury* were the medium.

Further to his fearlessness, Parkhurst was armed with an unorthodox and macabre sense of humor and perhaps this was what the ungodly dreaded the most in him. They could survive being proved cheats, hoods, shysters, burglars, double-crossers, chiselers, shakedown artists, or worse. But when he made fun of them and turned their machinations into a laughing stock there was little left for them to do but leave town.

At the moment of finding himself in a borrowed car halfway between Reckmeyer Field and Camauga, Parkhurst was forty, footloose and fancy free, having been divorced from an unsuitably literary wife who from her height of five-foot-three, had managed the considerable feat of looking down upon him because he was only a sportswriter. The marriage had been a catastrophe since it was based upon Victoria de Lacey's desire to have this big hunk of man and Parkhurst had fallen for the combination of her chic and intellect. Young, he had not suspected the penalties to be incurred by marrying a too-clever woman.

To begin with she had been a Lucy Stoner and kept to her maiden name. All the de Lacey's were *litterateurs* of one kind or another—novelists, playwrights, essayists. Victoria herself wrote articles for such high-domed periodicals as *The New Forum* and *Tomorrow's Day*. Inevitably when the sexual excitement engendered by being smothered by a giant had abated, Victoria had surrounded herself with a literary circle that regarded Parkhurst as a barbarian and a hack. He was never allowed to forget it. One day he had come home to find one of her clique's most erudite pansies reading a column of his aloud and all of them sniggering over some of his sentence structure and subject matter. Parkhurst had quietly shut the door upon the amusing scene, as quietly moved out of the house and got himself a lawyer. Victoria at least had had the good grace not to take any of his money. The experience had left Parkhurst bruised, lonely and wary.

The very noises that Parkhurst had been longing to hear now came drifting into his car window through the still air to forecast his approach to the fairground's center and soon, electric arc lights and gas flares were illuminating the night sky. Final confirmation was the cluster of gaily painted wagons drawn up on the outskirts of the exposition bearing the legend, 'MATSON'S MONSTER CARNIVAL'.

The jam-packed car park testified to the success of the Fair. Parkhurst paid his dollar entrance fee and soon was wandering contentedly and unrecognized through surroundings that were giving him the utmost pleasure. Here were the same stout wives gossiping over their entries of prize icing cakes, pies, tarts and preserves; gum-booted farmers, one leg up on the railing of pens, debating the finer points and qualities of 'hawgs'; cattle lowing, sheep complaining, tractor, plough and farm machinery booths with salesmen demonstrating under glaring lights and needlework, patchwork quilt and various home industry exhibitions along with last year's cotton and early tobacco leaf and other agricultural products of the State.

The Midway or amusement area, thanks to Matson's Monster Carnival, lived up to his fondest expectations. Here were the wheels of fortune, ring throwing and fishing games for huge dolls or ornate prizes, shooting booths, miniature auto racing and ski-ball and other games of skill and chance where the former was unavailing and the latter non-existent, set up on both sides of a wide and garishly illuminated street. Down the center paraded and pushed crowds of fun-seeking citizens, countrymen, shopkeepers, townspeople as well as those who had come from outlying districts. There was the heavenly noise of carousel music, clanging bells, whistles, cries of the barkers, and still more divine smells of sizzling hamburgers, molasses candy, popcorn, cigars and chewing gum.

Duke Parkhurst with a hamburger roll in one hand and a toasted waffle in the other, ambled along marvelling that what had looked like a dead loss could have turned into so magnificent an evening.

He paused before a booth where there was apparently some kind of a wrangle going on. The game was one in which various articles such as a small bedside clock, a pen and pencil set, an onyx ashtray, a pair of binoculars, a wristwatch and other objects of what seemed to be considerable value, were set up on square blocks of wood. For a dime one was given five rings to try to throw over an article. To win it, however, the ring must not only encircle the prize, but also settle down over the four corners of the square block.

Duke Parkhurst grinned fiendishly in the knowledge that four out of five rings handed the sucker would not fit over the wood. With the fifth, the pitchman demonstrated that they would. The odds against the player landing that one were astronomical.

The altercation involved three young men who had been playing and now, as their voices rose, one of them shouted, 'Goddam you, Ah said it was over and what Ah said goes!' And with that the man reached across the counter, snatched a pair of binoculars off the block of wood and with his two companions jerked away from the booth and was lost in the crowd. And now Parkhurst knew why Camauga had rung a bell. It was the home town of Lee Dockerty, Middleweight Champion of the World, and the fellow who had whipped away the binoculars was none other than Dockerty himself, out with two Cracker friends for a night of fun.

The incident had happened so quickly that Parkhurst had only a moment's vision of Dockerty in blue jeans and black leather windbreak with his two long-haired friends similarly clad and the impression that they had probably been drinking. He was glad that they had not seen him.

For a moment the columnist felt robbed of his carefree enjoyment of the adventure of the County Fair. To be reminded of Lee Dockerty and his manager Pinky Schwab was a source of irritation stemming from the fact that one of the few times that Parkhurst had been beaten on a good story was when *The Evening Post* had researched and printed a series alleging that the majority of Dockerty's build-up fights on the way to the championship had been crooked. The result had been a welter of threats, denials, counter charges but the point was that neither Dockerty nor Schwab had ever sued the *Post* for libel.

There was a further rumor that behind the scenes Dockerty was Mafia owned and operated by a shadowy *capo* known as Uncle Nono, a pseudonym used when he was mentioned just as in the days of Chicago prohibition it was considered both politic and healthier to refer to Capone as Al Brown. Uncle Nono was variously supposed to be a wealthy and highly respected Italian-American importer and socialite, a member of a gang of New York mobsters, or a Mulberry Street political boss. No one seemed to know and if they did, they were not telling. Parkhurst had to content himself with needling Pinky and Dockerty at every turn with innuendo and veiled references in the hopes of smoking out Uncle Nono or forcing some kind of break.

All this was over and above Parkhurst's personal dislike of Dockerty. In his book, Dockerty and Schwab were a discredit and a detriment to boxing.

The champion's presence in his home town of Camauga was not surprising. The coincidence was that Parkhurst should have been dumped there, too, and he wondered whether he could make something out of it for his column, and had to smile to himself. The Hoopla concessionaire, as the game was called, was a crook; he had been taken by another and had not had the guts to squawk. This returned Parkhurst to his previous good humor and then, putting the whole affair out of his mind, he continued his strolling.

His way led him past the usual girlie show with four skinny kids in fringed bikinis shaking themselves as a sample of more titillating sights to be enjoyed within; the freaks booth and oddities museum and others similar, none of which he considered worth his money, until he found himself standing before a brightly lighted front decorated with garishly colored six-sheets and posters announcing: 'EXTRA ADDED ATTRACTION! CAPTAIN BILLY BAKER, THE BERMONDSEY KID, AND HIS WORLD FAMOUS BOXING KANGAROO, MATILDA!'

Two pictures went with this text. The first showed a handsome stalwart young boxer with a British flag laced

48

through his gold-buckled, championship belt squared off in fighting pose against a giant kangaroo who had the Australian colors draped around his middle, in the center of a prize-ring, erect, gloved and menacing. The second olio depicted a boxer spread-eagled, flat on his back with a white-shirted referee counting over him and the kangaroo retired to a neutral corner, his arms raised, his gloved paws clasped over his head in the time-honoured gesture of victory.

At that moment the barker standing on the steps leading to the inside of the canvas-fronted enclosure was clanging a bell, holding a microphone to his lips and was spieling:

'Hurry, hurry, hurry! Step right up and see the world famous boxing kangaroo against Captain Billy Baker, The Bermondsey Kid, champeen of Great Britain in a contest of brains and brawn, skill and endurance . . . Bring the wife and the kiddies, one dollar admission, only one hundred cents to see man against jungle beast in a match to the finish . . . it's thrilling; it's chilling; it's sensational; it's educational . . . the science of self-defense as practised by the blood-sweating behemoths from the wild continent of Australia down under . . . You'd pay a hundred dollars to see a battle like this in Madison Square Garden in Noo Yawk, with no quarter asked or given and we're only asking you one easy buck . . . Thank you, sir . . . step right inside . . . And you, sir . . . bring your lady friend with you, she'll thank you for it. Hurry, hurry, hurry . . . the show's about to begin. . . .'

Parkhurst smiling happily to himself, said, 'That's for me,' extracted a dollar from his wallet, paid it over and went inside.

# VI

FOR ONCE THE outside talker had not been kidding and as Parkhurst entered and took a seat in the shadows of the back row of the small arena with rows of benches sloping upwards from the boxing ring in the center, one of not quite regulation size, the show was actually about to begin. The enclosure, with room for some four hundred or so spectators, was two-thirds full and with the arc lighting over the ring and the outer darkness gave the impression of a regular outdoor fight arena. The barker now appeared inside and by taking off his coat and tie and rolling up his shirt sleeves, metamorphosed into referee and announcer.

A bag punching platform had been set up alongside the ring and the announcer clarioned, 'Ladeez and gent'mun, to begin our performance introducing Captain Billy Baker, The Bermondsey Kid, former lightweight champion of England, in his super sensation and surprising symphony, a perpetual palpitation of punches on the light bag—Captain Baker!'

In the pool of light at the center of the little amphitheatre a stocky figure appeared clad in tights and boxing shorts, acknowledged the patter of applause with a wave and without further ado began to beat a rataplan on the inflated leather with such skill and changing rhythms that Parkhurst shifted his huge bulk on the narrow plank of the bench and sat up. This was not only professional, it was in a way beautiful. With his fists, elbows and sometimes even his head, he was playing music on the suspended bag.

But of course it was beautiful, Parkhurst thought. Baker was an old pro. He remembered the name now, but long

50

before Parkhurst's time, Baker had once gone fifteen rounds with Lou Ambers, then world title-holder, and had come close to toppling the champion from his throne. Faster and faster the bag rattled as Baker speeded up until it sounded like the rolling of a snare drum, when The Bermondsey Kid brought the exhibition to a close with one terrific left hook that tore the bag from its moorings and sent it out into the crowd who rewarded him with thunderous applause. They had already got some of their money's-worth.

From somewhere in Parkhurst's neighborhood came a voice, 'Ah bet you cain't do that, Lee.'

Another voice, 'Ah, cain, too. That ain't nothing. Yo gimmick the bag so it comes loose. Ah do thet at mah training camps. The boobs love it.'

Parkhurst located Lee Dockerty and his two pals sitting three rows in front of him.

Baker retired to a tunnel beneath the seats. The announcer took the ring again, waved his hat for silence and clarioned, 'And now, ladeez and gent'mun, the pugilistic sensation of the ages—Captain Billy Baker and his world famous boxing kangaroo, Matilda, from the wilds of Australia!'

The Bermondsey Kid came skipping out of the tunnel followed by a gray kangaroo, its progress seemingly made even more awkward by the fact that it had a pair of eight-ounce gloves laced to its forepaws. Baker had changed into boxing trunks with the Union Jack entwined as a belt and was likewise gloved. The kangaroo had a red rope tied around his middle. There was a startled murmur as Matilda entered the ring by bounding over the top rope.

Parkhurst found himself intrigued. What the hell was going to happen? Was that silly-looking beast really going to box with the old ex-champion, or was the whole thing a farce, another carney sell?

But what followed had an extraordinary snap and precision. The director of Matson's Monster Carnival, a canny showman, had apparently rewritten Billy Baker's act and produced it exactly as a regulation boxing contest would have appeared, except that one of the contestants was a kangaroo. Seconds attended Baker and Matilda in their corners. The announcer-referee made a fulsome introduction of both concluding with, 'May the best man or beast win.' The two received the usual instructions in the center of the ring, touched gloves, returned to their corners and at the clang of the timekeeper's bell, came out fighting.

They boxed two, three-minute rounds at the end of which at the final bell, the referee held up Baker's arm and Matilda's paw to indicate a draw decision and there was laughter and excited applause from the spectators.

High up on the rim of the arena, Duke Parkhurst was gripping the planking on which he sat with both hands and murmuring, 'Holy cats! I wouldn't have believed it if I hadn't seen it.' Nor was he certain that he did believe it and yet seen it he had—a kangaroo who could box like a man and better than most, an animal who appeared to have absorbed and was able to put into practice every trick and movement of ring craft.

'My God!' Parkhurst breathed to himself, 'that thing knew what it was doing!' and then immediately denied it. 'No, it couldn't be! It's just a damn clever act.'

When the applause had died away, Parkhurst heard Lee Dockerty say, 'Hell, that son-of-a-bitch couldn't fight his way out of a paper bag. That Limey wasn't hittin' him none.' One of his pals sniggered, 'Maybe he couldn't. Boy, he was movin'!'

Lee Dockerty said. 'Nuts! Ah seen plenny openings. Ah could fold that thing up like an acordeen with one punch in the belly. Come on, let's get outer here. Who's got the bottle? Ah wanna 'nother drink.'

The second pal produced a pint from his hip pocket and Dockerty took a good gurgle from it. 'Christ!' he said. 'What a robbery!'

They remained seated, however, for the referee was waving his arms again for silence. Billy Baker had retired from the ring. When there was quiet the referee announced, 'Ladeez and gent'mun, I'm now authorised on behalf of Matson's Monster Carnival to offer five hundred dollars in cash, paid on the spot to any man weighing no more than a hundred-and-seventy pounds who can stay two, three-minute rounds with Matilda.' And here he waved five greenbacks in the air, shouting, 'Here they are! Five one-hundred-dollar bills with Matson's Carnival accepting no responsibility for death or injury.'

There was a ripple of laughter, half amusement and half excitement, at the sight of the five green bills and people kept turning around and looking to the side and behind them as though to see if there was anybody in the enclosure who thought he could box well enough to have a go. There were no takers and then from the third row arose a grotesque figure in a battered silk hat, red fright wig and ragged clothes, waving his arms drunkenly.

Parkhurst's interest sagged. Beneath the red wig he had recognized as the stooge Billy Baker, who had done a quick change. This was the oldest circus gag, the supposed drunk in the audience who would take up a challenge and the show people played it to the hilt to the vast amusement of the audience. Baker stripped to a pink-and-white striped bathing suit, was duly weighed, made to sign a paper absolving the management from any responsibility. Then he went in with Matilda for a round of falls and knockabout tumbling which ended with Baker doing a back flip over the ropes.

Yet Parkhurst could not escape the feeling that all the while during the comedy act Matilda was in there trying—trying to tear Billy Baker's head off with his astonishing assortment of hooks and jabs. Apparently the kangaroo had no sense of humor and each time that the clown took a fall or a roll over, he would retire solemnly to a neutral corner and wait until Baker got up and came at him again. Boxing was evidently a serious matter to him and any time anyone approached him wearing a pair of those red leather pillows, he went out to destroy him. Actually the skill and the beauty of this part of the performance was all The Bermondsey Kid's, for he had to judge to a nicety just how close he dared come to one of Matilda's punches to enable him to take his fall.

Parkhurst wondered whether Matilda knew that he was being kidded as from time to time a bewildered expression seemed to gather about his muzzle as though he were aware that he had not connected properly and was wondering why his opponent had dropped. After this ridiculous thought Parkhurst shook himself and admonished inwardly, 'Come on, Duke, be yourself!' And he repeated, 'It's just a damn clever act.'

The announcer had one more try. Waving the cash he said, 'Last call! Anybody else? Come on, come one, come, all! Maybe you'll be lucky, Matilda has been known to miss.'

Parkhurst heard the pal who had the bottle say, "Whaddya say, Lee Baby? Take 'im on! With five hundred bucks we could all go downtown afterwards and get laid.'

The other chum coaxed likewise, 'Sure, Lee boy, you said you could take him with one punch in the belly. What about it?'

'Nah,' said Lee Dockerty, 'are you kiddin'? Do you think they'd let me in there, if they knowed who Ah was?'

The first pal said, 'Don't tell 'em. Give 'em your real name. Them carnival fellers ain't never heard of Ed Goordy. They don't know you're down here. They wouldn't recognize you. It would serve them right wouldn't it? Like that crook with that game you fixed.'

The second pal said, 'Boy, five hundred smackers for one punch! Annie May Lou's house would let us stay all night for that. Come on, Lee! Be a sport!'

'Okay,' said Dockerty, and the three stood up and waved to catch the attention of the announcer who shouted, 'Fine! Fine! I think we got someone here to take on Matilda. Come right down here, fellas.'

The three filed down to the ring. Parkhurst noticed that one of the pals was unsteady on his feet. He wondered how much Dockerty had been drinking. Whatever was going to happen, he wanted to be close to it. He made his way around the back row of the arena, descended on the opposite side and concealed himself in the shadow of the aisle of the tunnel.

The three were at the ringside now being interrogated by the barker-cum-referee-cum-announcer and Billy Baker. Parkhurst noticed that the attitudes of the showmen had stiffened to a sharp wariness. They were looking Dockerty over.

The barker asked. 'What's your name?'

'Ed Goordy.'

'Where do you live?'

'Four thirty-eight, East Chestnut.'

Baker asked. 'Have you ever boxed professionally?'

'Nah,' said Goordy/Dockerty. 'Do I look like it?'

Parkhurst noticed that he was standing slightly in the shadow, out of range of the ring lights which were not too strong anyway and he thought to himself: *He may be tight but he's still smart.*

'Boxed in the amateurs?' Baker asked.

'Nah!' replied Goordy/Dockerty. 'Just fooled around a bit with the gloves with the kids here. If I ever took up that racket, I'd git paid for it—like now.'

The barker asked, 'What do you mean, "like now"?'

Goordy/Dockerty replied, 'Like for five hundred bucks I can lick that thing there. You scared?'

The barker said, 'Just routine, sonny, just routine. Anybody here know you?'

Dockerty looked around and then pointed to a man sitting in the second row. He said, 'Yeah, he knows me.'

The barker turned to the man and said, 'Who are you?'

'Frank Lane, hardware merchant. I've known Ed. He's lived here since he was a kid. He's okay.' Somebody nearby sniggered.

In the covering darkness Parkhurst grinned to himself. Whatever, the home folks were not going to give Dockerty away. They knew who and what he was. Most of them in one way or another had been taken by the midway grift and would be delighted if Dockerty could get them even with the showmen.

'Okay,' said the barker, 'sign here,' and shoved a paper at Dockerty.

Lee said, 'What's that?'

'Nothing,' said the barker, 'it says we pay you five hundred bucks if you can stick two rounds with the 'roo or stop him, but we ain't responsible if you get hurt. That's fair enough, ain't it?'

'Yeah,' said Dockerty and signed.

Parkhurst said to himself, 'But not sober enough to read the fine print. Oh brother, Allah is great!'

'Okay, fella,' said the barker, 'take off your shirt and shoes and step on the scales.'

Dockerty complied. The barker looked at the meter and said, 'A hundred-and-sixty even. A pound your way. The 'roo comes in at a hundred-and-fifty-nine. Do you want to borrow a pair of shoes?'

Dockerty said, 'Nah, I don't need no shoes.' And suddenly suspicious he said, 'There ain't nothing funny about this, is there?'

'Nix,' said the barker. 'We're reputable people. You've got your own crowd here. You've seen the 'roo work out. It's your risk. Can your friends look after you in the corner?'

'Okay,' said Dockerty and climbed up into the ring and let them lace the gloves on his hands.

Parkhurst wondered whether Dockerty would be smart enough not to give himself away by the pro's habit of working the padding of the glove away from the knuckles while waiting for the bell. He refrained for Baker was eyeing him closely.

The columnist had a sudden rush of conscience. The show people had not caught on but he knew that Ed Goordy was Lee Dockerty, middleweight champion of the world and a cruel and heartless bastard. Ought he not warn them? The kangaroo was a valuable animal and obviously superbly trained; what if Dockerty were to destroy him or injure him permanently? It was a noble impulse and Parkhurst conquered it. Whatever happened it was not his funeral. He was a reporter and not bucking for a halo. Anyway, it would be flying in the face of Fate which had decided to deposit him as a witness to a once-in-a-lifetime happening. He retired a few feet further into the shadows.

Billy Baker went over to the bell, stop-watch in hand. Parkhurst thought he was looking anxious and wondered why. Had he, as an old pro, sensed that there was more to Mr Ed Goordy than just fooling around with the gloves with the town kids?

Matilda was still standing in the ring. Parkhurst wondered whether he was wrong, or whether he was really seeing a look of eagerness and excitement on the animal's face. It was so easy, particularly under the power of the suggestion of the circumstances, to read human thoughts and feelings into the expressions of dumb beasts. What was it called—anthropomorphism?

The crowd had settled into a tensed hush. The referee summoned Dockerty and Matilda to the center of the ring and gave them astonishingly clear and explicit instructions. 'Shake hands and come out fighting. No holding or hitting in the clinches. When I say break, break clean and step back. If you score a knock down, retire to a neutral corner during the count. That's all.' Then turning to Dockerty he said, 'Good luck, fella!' and held up the five hundred dollars again for everyone to see.

Dockerty and Matilda retired to their corners. Billy Baker clicked the stem of his stop-watch and pulled the lanyard of the bell.

# VII

THE FIRST OR 'green' edition of the *Daily Mercury,* printed on
Kelly green paper, is dumped in wire-bound bundles off the
circulation trucks at half-past eight the evening before the
morning of its date, often only half a newspaper, at least
sportswise, since the night baseball games are still going on,
prize fights incomplete and results from the Coast lacking.
These deficiencies are remedied by subsequent editions and
replates.

But the 'greenie', as it is known, is a newspaper complete
at least for its sports gossip and other columnists, editorials,
pictures and hoked up news stories and is eagerly awaited by
all the denizens of the Main Stem. There would always be a
small group collected around the series of news-stands
stretching from 39th to 52nd Street. There might be writers
and film people hoping to find a favorable review, press
agents anxious whether the item they had tried to plant
about their client in some column had made it, an avid
readership made up of actors, fighters, managers, hackies,
chorines, hookers, plain-clothes detectives and hustlers.

On the night of April 7th the circulation trucks as usual
whirled up Broadway, dumped their loads and charged off
again in and out amongst the theater-going taxis. Binding
wires were cut, eager hands plunked down coins, snatching
up the papers and then the owners thereof buried their noses
in the tabloid pages devoted to those departments which
concerned them most.

On the corner of Broadway and 44th Street, Pinky
Schwab, now encased in a fawn-colored top coat with velvet

collar, turned a deep purple shade under the lights of the Broadway spectacular signs flashing on and off, eyes threatening to burst from his sweating face as he emitted one squeal of outrage and thereafter became wholly inarticulate. His companion, Patrick Aloysius Ahearn, one cheek pouch extended, as usual, with shredded tobacco leaves, his leathery face excised of its usual expression of lugubrious cynicism, was howling with laughter, whooping, roaring, guffawing and slapping his sides. And three blocks north at 47th Street, Solomon Bimstein, every hair of his head standing on end, knees shaking, hands trembling with excitement, found his brain so bewilderingly assailed that he thought he would faint.

All three were reading the same article in the same paper on the lead sports page of the *Daily Mercury*. The forty-two point, black headlines across the top of the page blared:

'MATILDA OF AUSTRALIA NEW MIDDLEWEIGHT
CHAMPION! KNOCKS OUT TITLE-HOLDER LEE
DOCKERTY IN ONE MINUTE, THIRTY-SEVEN
SECONDS OF THE SECOND ROUND'

The story, datelined Camauga, Miss., April 7th, was 'by Duke Parkhurst', and began shrieking its news in ten-point type for the opening paragraphs, written in a style that only an old newspaper pro would have recognized as a subtle parody of the bombastic prose usually reserved by boxing writers for the description of the passing of a world championship.

'In an upset that will shake the world of Fistiana to its foundations, a new middleweight champion was crowned here last night in the Carnival Arena of the Mataposa County State Fair, when world title-holder Lee Dockerty was knocked unconscious and counted out after one minute and thirty-seven seconds of the second round.

'His conqueror and the new champion, a 159-pound, undefeated, six-foot kangaroo and Australian title-holder, gave Dockerty a boxing lesson in the first round, dropping him just before the bell, and knocked him cold in the second with a one-two combination of left and right to the jaw.

'It took ten minutes to bring Dockerty round after the knock-out and when he came to, he said, "What did they hit me with? Where did they go? Did they get my wallet?" He was under the impression that he had been ambushed in an alley and black-jacked with intent to rob. When he was told that he had forfeited his crown to the fightingest marsupial ever to reach these shores from Down Under, he refused to believe it and with his seconds leaped up and disappeared into the darkness, and this reporter has not been able to contact him since.

'But the facts of the débâcle and the circumstances under which the title changed hands are incontestable.

'Dockerty signed a contract involving a purse of five hundred dollars on a winner-take-all basis; he entered the ring at 160 pounds to his opponent's 159, thus laying his title on the line; an arbiter refereed the match under the rules prescribed by the late Marquess of Queensberry and there were nearly four hundred witnesses, including your reporter on the spot, who saw Dockerty, after absorbing a beating in the first round, knocked loose from his senses for the mandatory count of ten in the second. The short inside right that finished him off travelled no more than six inches. Dockerty fell as though drilled by a high velocity bullet and never stirred.

'It was under his real name of Goordy that Dockerty accepted the challenge of the management on behalf of Matilda, Australian champion boxing kangaroo who had been giving a series of scientific exhibitions in the art of self-defense in conjunction with his trainer and sparring partner, Billy Baker, The Bermondsey Kid, ex-lightweight champion of Britain.

'All the regulations pertaining to the proper conduct of the bout were complied with. Dockerty had two friends with him who served him as seconds in his corner.

'After receiving their instructions in mid ring from the referee, the two contestants touched gloves and at the bell, came out fighting. And herewith your eyewitness offers a blow-by-blow description of probably the strangest championship fight in the annals of boxing:

'*Round One:* At the bell erect on his hindlegs, Matilda

was at center ring with one hop. Lee Dockerty met him there with his own rush and uncorked a left hook to the belly that was supposed to end the fight without any further exertion on his part except that Matilda was not there to receive it, having apparently anticipated and pivoted away from it. The force of Dockerty's swing carried him forwards so that he almost fell out of the ring. Some of the spectators, under the impression that he was some kind of comedy stooge, gave the effort a big laugh.

'The kangaroo refrained from hitting Dockerty while his back was turned, a sportsman-like gesture which drew applause. Dockerty was red-faced with fury. One of his seconds called out, "Settle down, Ed baby. You've got plenty of time."

'Matilda's boxing stance was loose and elegant, left half extended, right hand high and covering his chin, with an expression of interested concentration on his face.

'Dockerty tested him with a straight left jab. Matilda blocked it with his right, leaned in and simultaneously cuffed Lee with three short left hooks and a right to the head. In the opinion of this observer, if Matilda had put any sting behind the punches, he might have knocked out Dockerty then and there. Instead he stepped back as though to study the effect.

'Dockerty shook his head to clear it, tucked his chin deeper into his left shoulder and shuffled forward more cautiously. He looked fit to be tied. The champion threw two left jabs and a looping right cross aimed for Matilda's eye. Matilda blocked the jabs with his right glove, knocked up Dockerty's right with his own left, clubbed the champion on the side of his head, reddened his face with three lightning lefts of his own, sent him spinning into the ropes with a right that glanced off Dockerty's shoulder. Matilda followed him up trapping the champion and sending two lefts and rights into his stomach, doubling him over, then straightening him up with a right uppercut. Dockerty fell into a clinch, hanging on for dear life. Matilda extended his long dark muzzle, a pink tongue emerged therefrom and lovingly caressed the side of Dockerty's face.

'The champion protested to the referee, "Hey, make him stop licking me!"

'The referee said, "That's how he shows he likes you."

'Dockerty snarled, "Well, I don't like him. He stinks! Tell him to cut it out!"

'The referee said, "Break, Matilda!" and the kangaroo stepped back, looking a little hurt I thought, but it may have been my imagination. For the next minute he contented himself with avoiding Dockerty's attack by a magnificent exhibition of sidestepping and blocking with occasional moments of feinting the champion dizzy, but refraining from taking advantage of the openings. He appeared to be enjoying himself.

'Matilda was wearing a red rope around his midriff, apparently to mark the foul line, but a full-grown kangaroo, when standing erect, carries his prized possessions somewhat lower than those of a human being, say on a level with his kneecaps. In a seething rage, Dockerty aimed a looping left at these which missed only by inches as Matilda pivoted back on his tail. The punishment was immediately forthcoming and before Dockerty could recover from the miss, Matilda clubbed a short left hook to the temple and Dockerty sat down hard just before the bell. MATILDA'S ROUND.

'Dockerty's two seconds jumped into the ring to help up the fallen gladiator, one of them saying, "Say, Ed, maybe you'd better get outer this."

'But Dockerty was already on his feet cursing, "Get away from me, you bastards! I'm all right. I'll kill that son-of-a-bitch next round!"

'Matilda retired to his corner but also rejected sitting down and stood with his arms along the ropes, not even breathing heavily, while his second gently ruffled the fur on top of his head.

'*Round Two:* At the bell Dockerty was off his stool with a rush. He had apparently decided there was no profit in trying to take Matilda at long range or box with him, and he now tried to outslug and overwhelm him with a whirlwind of punches thrown from all angles. With only another few minutes to collect the purse and restore his injured pride in front of the home crowd, who were now sitting in stony and startled silence, there was no need to hold back.

'Like any experienced boxer, Matilda got on his bicycle, backing away from the flurry, letting Dockerty expend himself, never taking his dark watchful eyes off his opponent and feeling apparently either by instinct or osmosis when he was in danger of being trapped in a corner or against the ropes. Matilda moved in smothering Dockerty's attack in a powerful holding manoeuvre in which, by resting both gloved paws on Dockerty's arms in a heavy clinch, he held the struggling champion powerless.

'Matilda now licked the left side of Dockerty's face, then the right and finished with a passionate buss full on the mouth. I noticed that the little Cockney with the bell and stop-watch was grinning from ear to ear. Later on I learned that what Matilda was bestowing upon Dockerty was the kiss of death. He had had his fun; the performance was about to end.

'The referee ordered, "Break, Matilda!" and as the kangaroo stepped back with his paws down, Dockerty, instead of doing likewise, put everything he had into an illegal right to the jaw.

'With an almost snake-like movement, Matilda reversed his direction, moved his head just inside the blow so that it curled around his neck and, experienced an observer as I am, I must confess I did not actually see the knock-out punches beyond two twitches of Matilda's shoulders which must have powered a lightning left hook and an even faster right cross. But I heard them; the crack of the first, the thud of the second and the crash as Dockerty went down. When they fall forward onto their faces you know they're not going to get up.

'Matilda hopped to a neutral corner, shook himself once, then raised his gloved paws above his head even before the count of ten was finished. He knew it was over. Matilda, the winner and new champion by a knock-out.

'And so,' summed up Parkhurst, 'yet another boxing king passes into the limbo of ex-champions, nor will he be greatly missed. Dockerty, who held the title only briefly and defending it but once in two years when he knocked out Spider Ray, was not to be compared with such greats as Stanley Ketchel, Harry Greb or Ray Robinson, nor was he too popular during his reign. But he was courageous and had a punch which made him always dangerous.

'The loss of his crown to Matilda of Australia could be

63

laid somewhat to overconfidence on the part of Dockerty who, as well, did not seem to be at the height of his form tonight, but mostly to the fact that his opponent was his master at every angle of the game. It is safe to say that during the four minutes and thirty-seven seconds of fighting, Dockerty was unable to land a blow heavier than a graze upon Matilda.

'Who there is around at the moment capable of challenging the new middleweight king is difficult to say. The current crop of middleweights is not too distinguished and any aspirant who enters the ring against Matilda will want more skill and experience, not to mention a more punch-proof chin, than is ordinary.

'Pinky Schwab, Dockerty's manager, was not present to witness his fighter's downfall, but will undoubtedly press for a return match and the gate that such an attraction would draw. However, like all new title-holders, Matilda will probably wish to cash in on his success with a few exhibitions or a defense against one or two of the lesser challengers before taking on Dockerty again.

'"*Eheu fugaces . . . labuntur anni*", so passes the old order, "The King is dead! Long live the King!" This writer hails the new middleweight champion from Australia, a boxer, a fighter, a puncher, a gentleman and a kangaroo.'

# VIII

BEGINNING WITH THE corner of Broadway and 45th Street where Patrick Aloysius Ahearn and Pinky Schwab were practically drowning out the noise of Times Square traffic, the former with peals of sardonic laughter, the latter with roars of rage, Duke Parkhurst's bombshell was doing its work. Practically inarticulate with fury, his hat on the back of his head, the sweat pouring from his fat red face, Pinky was shouting, 'Jesus J. Christ! What's that stupid son-of-a-bitch Lee gone and done now? He just said he was going back home for a few days. . . .'

Patrick Aloysius had recovered his breath long enough to say, 'Got himself pissed and knocked out by a goddam kangaroo, and Parkhurst is having a field day. Brother, if you know what's good for you, you'll leave town for a couple of days.'

Pinky had been thrown into such a state of confusion by the story that he had to refer to it again. 'But it wasn't no title fight,' he said. 'Lee didn't have no match scheduled with nobody. What the hell does Parkhurst mean, Lee ain't the middleweight champion any more? I can sue him for that.'

Patrick Aloysius's face was a study in delighted malice. 'Yeah, that's right,' he said, 'just you go and sue him and finish the job. Looks like your boy went after some easy money, forgot to duck and somehow Duke Parkhurst was on the spot to see it. Bring that out in court.'

Pinky, who by now was reading the story for the third time and was beginning to be aware of the preposterousness of it, was saying, 'He's kiddin', ain't he?'

Patrick Aloysius saw no reason to withdraw the needle, shifted his chaw and said, 'Duke Parkhurst don't kid.'

Pinky was almost crying now. He said, 'But it's got to be a gag, don't it? You can't lose a championship over a two-round exhibition. You gotta have fifteen rounds, a contract, a licence from the Licence Commissioner and everything, don't you?'

'Yeah? Who said so?'

Pinky stammered, 'Well, like, the law, ain't it? There's got to be a law. He's just trying to be funny.'

'Uncle Nono won't think it's so funny,' Ahearn said, putting in the final stinger. 'Wait until he reads this. He just loves to be made a fool out of.'

The color of anger was now drained from Pinky's face and replaced by the pasty pallor of fear and he quavered, 'But he wouldn't believe that. It couldn't have happened. . . .'

Paddy Aloysius flicked the article with the back of his hand and said, 'Duke Parkhurst's put his name to it.'

Pinky blurted, 'I gotta get me to a phone,' and stumbled off, leaving Ahearn alone on the corner to reread the story and mutter, 'Now I wonder what really did happen?' A sudden memory dawned and he muttered, 'Matson's Monster Carnival—My God, that must be Bimmie's kanagaroo, the one with the hell of a punch.'

Two blocks north where the merged thoroughfares of Broadway and Seventh Avenue part company, that individual, young Solomon Bimstein, his knees still shaking, was entertaining visions of riches heretofore unimaginable: whole wardrobes of handmade suits, shirts, shoes and ties, town house, country house, his and her motor-cars, private jet, and Hannah Lebensraum now Mrs Sol Bimstein, lopsided with jewelry, all but invisible in draperies of mink, queening it over Upper West Avenue.

For if one could believe what Duke Parkhurst had written he, Bimmie, was now manager of the new middleweight champion of the world. The sounds that came bubbling from Bimmie's quivering lips, could they have been identified, were a repeated litany, 'Oh boy, I gotta contrack! I got the control! It says fifty-one percent me, forty-nine percent, Baker!' In his desk drawer was the paper giving him a one percent majority ownership in the act and everything pertaining thereto, including Matilda.

66

Bimmie glanced at the electric clock ticking off the minutes on the old Times building and asked himself, 'Who is the smartest little Jew on Broadway at eight-forty-seven tonight?' And there was only one answer. He wavered between a Cadillac, a Lincoln, a Mercedes and a Jaguar and decided he might have all four. He read the story again. There it was, under Duke Parkhurst's signature, and what Parkhurst wrote and said was Gospel. Bimmie was nearly killed by a taxicab as he raced across the street to a drugstore to call Hannah. What would Mr Lebensraum say when he heard his daughter was going to marry the discoverer and manager of the new middleweight champion of the world?

The green edition appeared not only upon Broadway to wreak its effect but was also immediately laid upon the desks of the sports editors of *The Times, The Post, Newsday* and the *Daily News,* where it was grabbed by them and their boxing writers for a quick look over to see what stories, if any, on which they might have been beaten.

At the *News,* Jack Horne said, 'What the hell's got into Duke Parkhurst? Did he get drunk?'

His editor was not so certain, and said, 'Would they have printed it if they thought Parkhurst was plastered? I never heard of Duke drinking.'

Gill Hobart at *Newsday* said, 'It's got to be a rib, hasn't it?'

*His* editor said, 'How can you tell with Parkhurst?'

Hobart queried, 'What do we do?'

The editor said, 'Nothing for the moment.'

On *The Post,* Frank Petrie asked, 'Ought I go down there? Supposing it really did happen?'

His boss said, 'No. We'd better wait. But get hold of Pinky Schwab and see what he has to say.'

The dignified sports editor of the dignified *Times* asked of his dignified boxing writer Cassius Jones, 'What do you think?'

Jones said, 'If Duke Parkhurst said that's what happened, that's what happened.'

The editor nodded and said, 'We'd better cover ourselves. Run a paragraph as though it was an A.P. story, but use it "as alleged".'

Alone in his office at the *Mercury,* Justus Clay the bald

diminutive managing editor who had a face like a monkey's but a tongue like an asp's, permitted himself a grin, since it was his policy never to smile in public—at least not on *Mercury* premises. He then pressed a button for his secretary and dictated a wire: 'DUKE PARKHURST C/O WESTERN UNION CAMAUGA MISS. ITS A GAS THE WHOLE TOWNS LAUGHING STOP IF TRUE WIRE PICTURES OF KANGAROO WITH SECOND-DAY STORY STOP CLAY.'

The secretary said, 'Mr Parkhurst's second day story is just beginning to come in now, Mr Clay.'

The editor said, 'Okay, let me have it in takes.'

Simultaneous with the arrival of the half sheet of Parkhurst's copy from the wire room, the studio sent in prints of wire photos which had come in. Parkhurst was too old a hand not to accompany his story with pictures were there any available. Unfortunately there had been no photographers at the ringside to record the flattening of Lee Dockerty but Parkhurst had sent some copies of postcards of Billy Baker with his famous boxing kangaroo Matilda, and also some prints of Baker in action with the animal which he had arranged the following morning with a local photographer.

Shaking his head, Clay studied the big animal squared off against The Bermondsey Kid and the expression of pride, interest and truculence upon its features, and in particular one shot where the trainer had ducked under a left hook, and such was the speed of the punch that the camera had been unable to stop it and the kangaroo's left arm was a blur.

'It couldn't have happened,' Clay murmured to himself, 'but by God, if Parkhurst said it did, then it did. Look at that bastard!'

Alone once more, his malevolent smile cracked his features and he set himself to reading paragraphs at random from Parkhurst's second-day story. A brief A.P. despatch about the Southern Airlines Flight 307, New York to San Antonio Boeing being forced down at Reckmeyer Field had told him Parkhurst had come to be there.

'Mr Baker,' so ran one of the takes of Parkhurst's follow-up piece, 'explained Matilda's extraordinary ambivalence to me. He boxes with a heart full of love for his opponents and shows this by the affectionate licks he bestows upon them in the clinches. It seems that such is Matilda's interest and delight in his skill that any adversary able to extend him, affording the opportunity of the full display of his art, becomes at the same time enshrined in his kangaroo affection. Thus he fights beset neither by rancor nor cruelty but with a combination of compassion, gratitude and love. It is when Matilda has judged that his foe has shown him all he has and has nothing further to offer in the line of novelty or excitement that this emotion of thankfulness reaches its full flower and in the last clinch Matilda lavishes one final, endearing kiss upon the countenance of his victim before zapping him for keeps.

'Your correspondent witnessed an example of this when in the last embrace just before dropping Dockerty for the count, Matilda deposited a moist and loving buss full on Lee's mouth. His trainer refers to it as the Kiss of Death and prepares to sweep up the remains of its recipient.'

The editor scribbled a sub-heading, 'Don't Kiss Me Again', over the paragraph and buzzed for the copy boy to come and take it away. He then resumed his reading with a snort of, 'Oh no!' as he came upon Parkhurst's utterly shameless descent into the type of hagiography used to make readers aware of seeds of sainthood in a character. 'Mr Baker narrated an incident that occurred when Matilda was approximately sixteen months old and they were travelling with the Ralston and Fotheringay Circus, where the ex-boxer was giving exhibitions of bag punching. The Circus zoo owned two further kangaroos, an old gray male weighing some 175 pounds, and a younger animal. One day when the three beasts had been turned into a field to graze and enjoy the open air, the big gray, who in the wilds might have been considered the 'Old Man' of the herd, began to abuse the younger one. Kangaroos in their natural state "box" one another; that is rear up on their hindlegs and strike and slash with their front paws to establish supremacy in the herd and choice of mates.

'Worried that his own animal might be the next to suffer from this aggression, Baker hurried to prevent it when to his amazement, he observed Matilda approach the bully in defense of the abused 'roo and challenge him directly.

'Hardly able to credit his eyes, the ex-lightweight champion saw Matilda smartly stop the enraged rush of the "Old Man" with a straight left to the muzzle, followed immediately by a lightning left hook to the body which caused his attacker to drop his guard. Matilda then moved in whipping over a short right to the chin of the big 'roo, knocking him unconscious.

'It was through this act of courage and gallantry that Mr Baker first became aware of Matilda's extraordinary natural talent for the finer points of the Manly Art.

'Then began a series of daily lessons with the result that by the end of a year, Matilda had become a master boxer with a passionate love of the game for the game's sake, as well as exhibiting all the attributes of a sportsman and a gentleman.'

But the part of the story that really put the cat amongst the pigeons when it appeared on the street that night and was reread in the office of Pinky, Patrick Aloysius and Bimmie the following morning was Parkhurst's concluding remarks.

'Queried about the future of the new middleweight king, Mr Baker said, "We'll be a fighting champion. Matilda has never barred anyone up to ten pounds over his weight and we'll continue to take on all-comers." Asked whether he would give Lee Dockerty a return match, Mr Baker said, "Of course we will, provided he proves himself to be the most worthy challenger. I think there ought to be a series of elimination contests to determine the best man. If Dockerty comes out on top, we'll meet him anywhere, any time, though I understand there are a lot of other good boys about."'

Ahearn, his expression grave, was reading the article as though it were Scripture, and said with a perfectly straight face, 'Very well put, if I may suggest. This man Baker seems to know his business. I have been saying that we ought to have a middleweight elimination series to make Dockerty lay his title on the line. Christ, he wouldn't even fight that bum Jimmy Cardo over the weight limit! I've got a young kid coming up who could make plenty of trouble for anybody.'

Pinky exploded, as he was intended to do, 'For Chrissakes, Patrick Aloysius, cut it out! There ain't gonna be no elimination. Dockerty ain't fightin' nobody. Who does Parkhurst think he's taking for a ride?'

Bimmie, still on cloud nine, said, 'Not nobody. Your boy stuck his neck out and got it crowned. I'm the manager of the new middleweight champion of the world.' Something on the page caught his eye, for he read it again and then said, 'Hey, what's all this crap about Baker saying we'll take on all-comers? Is he crazy? I'd better get on down there before he shoots his mouth off any more. We don't have to fight nobody we don't want to. We're the champ now.'

Patrick Aloysius said darkly, 'That's the boy, Bimmie. You're talking like a real manager already.'

Pinky's detonation reached a new volume in decibels, 'You, you little Jew bastard! You ain't nothin'! You got no manager's licence; you got no second's licence; you don't know your elbow from a hole in the ground! You're gonna get your ass out of this office! Maybe you put Parkhurst up to trying to make a monkey out of me and Lee.'

Bimmie got off his chair, his hair ruffling like that of a cockatoo, his fists balled threateningly, 'Who are you calling a Jew bastard? What are you? A goy suddenly, from Park Avenue?'

Pinky paled slightly and half apologized, 'Maybe I shouldn't of called you a bastard, but I got no use for a snot nose, even if he's a Yid.' And thereafter the catastrophe overwhelmed him again and turning to Ahearn he said, 'What was Parkhurst doing down there? How did he get there? Why did he go? How did he know that Lee was going to...' Here something told him that he might be getting in deep and he suddenly clammed up.

Ahearn let the opposite side of his mouth from his tobacco chaw slope into a grin. He said, 'You've got all the luck, Pinky. There was a piece in the paper said Parkhurst's plane was forced down just in time to let him see your boy get knocked ass-over-teakettle.'

'Sure, that's right,' Bimmie said. 'He was there. He seen it.'

Pinky blew again, 'Any more out of you and you'll get a bust in the mouth! Git outer here!'

Patrick Aloysius eyed him coldly. He said, 'Careful, Pinky. You ain't in shape. If you ever put your hands up you'd probably have a heart attack. Anyway, I got as much say as you as to who stays in this office.' He plunged the needle again, 'How about selling me a piece of your champ, Bimmie? I'll give you twenty grand for ten percent. You may be needing some help.' But he was only half serious.

Bimmie said, 'Are you kiddin'? What do I want to sell for, just when I got it made? I got fifty-one percent of the world's champ. You seen him in the gymanasium—I guess that wasn't no ack after all. Like Baker told me, he's a natural. We'll clean up: champeen's end of the purse, television, movin' pictures....'

Pinky's stricken squeal penetrated throughout the entire floor of the building, 'What are you trying to do—drive me out of my mind? You ain't got no claim to my title! That wasn't any regulation fight!' He was suddenly struck by an idea and his round face went apoplectic red again, his hands were shaking as he plucked the telephone from its cradle, dialled and practically jammed the speaking part of the instrument to his mouth. He said, 'Hallo, long distance? Operator? I want the Chairman of the Boxing Commission in...' He looked around desperately at both men for a moment, 'What's the capital of Mississippi?'

Ahearn's jaws were working on his cud rhythmically and without a break in their movement he replied, 'Jackson.'

'...Jackson, Mississippi,' Pinky was shouting. 'No, I don't know the number...Jes' ask the operator down there for the Chairman of the Boxing Commission and shake it up!'

He sat there strangling the neck of the telephone, a wild glare in his piggy eyes, as he contemplated the other two men and their coming deflation.

The voice of the operator returned to the instrument. Pinky said, 'What? What? You're sure? But there's got to be!' The color was draining from his face and he said once more weakly, 'You're sure? Okay, skip it,' and hung up. He looked dazed. He said, 'She said there ain't no Boxing Commission in Mississippi.'

Ahearn said, 'I could have saved you that call, Pinky. Mississippi hasn't got a State boxing law. They got local option. I doubt whether a dump like Camauga got any boxing commission.'

72

Bimmie leaped upon this revelation. 'So what Mr Pockhurst says goes. If there ain't no boxing law, and no Boxing Commission and Lee went in there and got put in the deep freeze by my kanagaroo, then that's it, ain't it, Patrick Aloysius?'

Ahearn, the cagey old fight manager and veteran of a thousand odd situations coming up in the boxing game, who up to this moment had purely been enjoying needling Pinky in his discomfiture but considered Parkhurst's report no more than a gigantic joke or even possibly a hoax, was suddenly looking thoughtful. What if what Bimmie was saying was true? In a State where there was no boxing law, Boxing Commission, rules or regulations and the champion at weight, a title fight could be one or, like in the old days of the English prize ring, a hundred rounds, depending upon what agreement had been signed prior to the match. If everything that Parkhurst had written was a hundred percent true, Bimmie's 'kanagaroo' might yet prove a gold mine, and Patrick Aloysius was determined to get in on the ground floor. He said, 'You could have something there, if Parkhurst doesn't let up. He's syndicated all over the country.'

Bimmie jumped up, clapped his hat on his head and made for the door.

Ahearn said, 'Where are you going?'

Bimmie answered, 'Down to the Licensing Commissioner to apply for a manager's licence, like what Pinky said I ain't got.'

Patrick Aloysius reached for his skimmer. He said, 'I'll come along with you, son. You might need someone to stand up for you.'

Left alone in the office, a half insane glare still clouding his eyes, Pinky reached for the telephone again and the long distance operator, shouting, 'Get me Lee Dockerty in Camauga, Mississippi!

The difficulty of regulating the boxing law with a three-man commission which could never agree had recently led the State of New York to abolish the triumvirate and invest all power in a single Licensing Commissioner responsible directly to Albany. This might have been beneficial to the situation if the State Legislature had been interested in boxing, which it was not, and the Licensing Commissioner had been other than a political appointee, a ward heeler by the name of Colonel Wild Bill Wildman, foisted upon the city by a boxing lobby concerned only in having a free hand, practically, in gaining its own ends.

The Colonel, who originally hailed from Arizona, was a town character who still affected high-heeled cowboy boots and had a licence to pack a six-shooter. He had acquired his military title through symbolic attachment to the staff of a bourbon-soaking, Southern Governor.

He wore a toupée that was always slightly askew on his head and which Parkhurst, in a moment of pique at some ruling of his, once wrote covered the brains of a not yet fully-fledged marmoset. This, however, was a libel which the Colonel disdained to pursue for he was actually a shrewd politician and, if nothing else, when the chips were down he could deliver the votes.

This was the person, then, who sat closeted with Bimmie and Patrick Aloysius Ahearn in his downtown office, his feet up on the desk, his coat off enabling them to see the six-gun and cartridge belt strapped around his middle. His wig was three degrees off center and he was trying desperately to grasp at clues as to what was going on, but was still about six furlongs behind the pace.

To the intense disgust of Patrick Aloysius, Bimmie had launched into a narration of the story of Matilda beginning with Mr Matson's Carnival, progressing through the visit of Billy Baker, a recitation of Duke Parkhurst's columns which he knew by heart, and winding up without apparently the drawing of a breath, '...and like Duke Pockhurst said, Lee Dockerty is down there at this fair, looking for some easy money and my kanagaroo flattens him and so I got to have a ... Ow!'

This exclamation of pain was occasioned by the fact that Ahearn had had enough and, without a change of expression, had trodden hard upon Bimmie's foot while at the same time scoring a dead center in the gaboon which the Colonel kept at the side of his desk for visitors.

Wild Bill fuffed and sputtered, 'What? What? What are

you talking about? I don't get it. Who's this woman Matilda? What's she got to do with Lee Dockerty? What's all this stuff about a kangaroo? I read something in Duke Parkhurst's column. What's all that got to do with me?'

Patrick Aloysius, having prepared himself for speech by emptying his right cheek cavity, spoke up from that side of his mouth and said, 'Nothing, Mr Commissioner, nothing at all. You know how Parkhurst is, he likes to kid around a lot in his column... Bimmie here's young and kind of gets carried away sometimes, but he's a good boy, Mr Commissioner. I need some help managing my stable of fighters. I've offered Bimmie the job to look after some of my boys when I have to go out of town, or give me a hand around the office, but maybe if he's called upon to work in a club sometime, he ought to have a manager's licence.'

The Colonel was at once grateful to Ahearn for providing him with something that he could grasp immediately. 'Oh, I see,' he said, 'a manager's licence, eh? I'll have to look into his record.'

Ahearn said, 'He hasn't any record, Mr Commissioner, he's a clean-living Jewish boy from a respectable Jewish family. I can get you plenty of references. He's never been in jail or had any trouble with the police. I've known him for five years.'

'That's nice,' said the Colonel. 'What's his name?'

Ahearn said, 'This is him, here—Sol Bimstein; 1620 Seventh Avenue.'

The Colonel took his feet off the desk and inspected Bimmie. He said, 'He can sure talk a lot. I guess you need that in your business. Are you standing up for him?'

Ahearn said, 'Yup. He's been in our office—an agent for night club acts.'

Assailed as he was by the many doubtful characters around the prize ring, Wildman was inclined to put his faith in reliable oldtimers whom he knew. Ahearn had been in the boxing game since the Walker Law had been enacted. Besides which, in the Commissioner's opinion there were far too many Dagos intruding into the sport. The Irish and the Jews always got along. He said, 'He looks like a nice boy. I guess there wouldn't be any trouble, Patrick Aloysius.'

Ahearn did not even bother to wink at Bimmie as he said, 'And while you're about it, Mr Commissioner, I want to apply for a second's licence for a new man I've taken on. Make it out in the name of William Baker, same address as Bimstein, here. He's great on cuts and I know how you don't like a lot of bleeding in the ring, with which I'm compelled to agree with you, now that so many ladies attend our soirées.'

The Colonel made a note, happy that something so simple and cut and dried had come across his desk and therefore intending to implement it immediately. Bimmie and The Bermondsey Kid were practically licenced. He said, 'Thank you, Mr Commissioner, you won't regret this,' jerked his head at Bimmie in a gesture which said, 'Come on, let's get out of here while the getting's good.'

'Boy,' said Bimmie in admiration, 'you put that over!'

'Christ!' Ahearn said disgustedly, 'you and all that crap about that kangaroo and Duke Parkhurst and a new middleweight champion! Another minute of that and he'd have thrown us both out of there. Now look, what about letting me in for ten percent of your fighter? I'll make it twenty-five grand.'

'Nah,' replied Bimmie, who was not burdened ever by any recourse to such a thing as gratitude. 'I'm manager of the new middleweight champion of the world, I got a second's licence for Billy Baker and we're gonna clean up.'

Ahearn was neither surprised nor disappointed, since gratitude was something one did not expect in the fight business. He said, 'What are you going to do?'

Bimmie said, 'Get on down there and take over before The Kid shoots his mouth off too much. We ought to be able to gross a grand a night for exhibitions.'

Patrick Aloysius nodded. He said, 'I'll come along with you. You're new at the game. You might be needing some more help.'

Bimmie made no objection. As long as Ahearn was working for free, he could be useful.

# IX

MATSON'S MONSTER CARNIVAL had shaken the dust of Camauga, Mississippi from its gilded wagons and, trekking across the border into Oklahoma, was now entertaining the citizens of Kiowa.

The news of Matilda having knocked out the middle-weight champion of the world incognito had speeded over the nation via the wire services and syndication and Captain Baker and his boxing kangaroo were doing a land office business. Mr Matson had raised Billy Baker's salary and promoted his show to the place of stardom in the traveling Carnival. Yet withal, the little Cockney was not too happy.

The Bermondsey Kid was a loner who for many years had been far away from home. He had no family, his wife had died in Australia and always on his mind had been to return to the wonderful, noisy, sooty Thames-side neighborhood, buy back the pub his grandfather had once owned and there serve behind the bar to customers who would be drawn by his reputation and recollections of his glorious past. They would remember Billy Baker in Bermondsey.

He had worked hard for this objective but for one reason or another had never been able to save any money. Periods of work were followed by lean weeks between jobs which ate up his capital, or rather Matilda devoured whatever he had been able to put by. Circus pay was not high; the big score had always eluded him. His old age spent in the warmth of his own cosy public house became a fading hope.

And then suddenly, what for so long had seemed like an unattainable dream loomed close to realization. He was a star feature now and a national figure. There had already been enquiries for television appearances. Yet, in spite of his unexpected turn of fortune's wheel, all was not beer and skittles, as they used to say down by the river. For Baker truly loved the strange animal who had been his support for so long and the prominence he had acquired overnight brought problems along with fame.

Up to that time Matilda, buttressed by his extraordinary skill, had seemed in no danger from the occasional awkward clodhopper or even ex-amateur who crawled through the ropes in the hopes of collecting some easy money. Matilda could take care of that type seven to the half dozen. But now with all this publicity and hullabaloo the competition was likely to become more critical. Baker knew his four-footed friend inside and out and if he were to suffer an injury, he would never forgive himself. With a tyro in the ring coming up against Matilda's skill there was never even any question of landing a lucky punch. But no more than a fair to middling pro might reasonably be expected at least once in a round to get through even the most perfect defense. The thought made The Bermondsey Kid shudder.

Furthermore, a stranger in a strange land who spoke in what practically amounted to a foreign language, friendless, he had no one to talk to, to discuss his problems and ask for help, or adivce. Thus, to compensate for this need for conversation, even though he knew it was useless, he would address Matilda while brushing and grooming him in his stall, reflecting his worries.

'You know what you've done, Matilda? You've zapped the champion of the world and knocked him arse over tit, and the 'ole blinkin' world knows about it. So yer can't fool around any more, them days is gorn forever. From now on you've got to be sharp every minute you're in there. You've had yer fun and yer can play around with me, but you can't no more with them other blokes. I know yer likes ter see what yer can do, but if yer gets an opening now, you zap him quick! I'm tellin' you, Matilda, I'm an old hand. They used ter call me "Lightnin'", I was that fast, but yer can't stop 'em all, as witness me bashed-in beezer and other marks, the result of a moment's carelessness 'ere and there. Mind you, I

78

was never the man you was, or had one tenth of your speed and savvy, but yer can't afford to take chances any more. We'll try to keep the old pros out if we can spot 'em but you never can tell when one could slip through, like that barstid Dockerty who was in there tryin' to do you a mischief. So from now on, the first time you sees an opening, yer lets 'im 'ave it. Get it?'

Matilda was lying on his side, on one elbow, his head braced upon his forepaw. He loved being chatted up by Billy Baker and was making contented 'uck-uck-uck-uck-uck' noises in his throat. The kangaroo, of course, did not understand a word, which was a pity, for had he known what the future held in store for him, that shortly he was to be allowed to zap real professional fighters at the rate of one or two every month, he would have been the happiest marsupial ever to come out of Down Under.

But the problem of this manner of communication was insurmountable. What was a kind of substitute for it was the perfect trust and affection the animal had for Billy Baker who fed him, groomed him, petted and loved him and every so often provided him with entertainment by turning him loose in a boxing ring with another kangaroo—for he saw the whole world of humans merely as different species of kangaroo—all of them challenging him and thus, as noted, furnishing him with a few moments of fun until they made the inevitable mistake.

A roustabout came to the door of the stall and said, 'Hey, Bill! There's two gents here to see you.'

Baker gave Matilda a pat on the flank and went out to find Bimmie and Patrick Aloysius, who had caught up eventually with the carnival in a rented car.

Alone with them in their hotel room he let out his ire:

'Look 'ere, what are you fellows trying to do—ruin me kangaroo? What's all this balls about 'im being middleweight champion of the world? And who's responsible for sending that there champion bloke in with my Matilda? Taking advantage of a dumb animal, is what I calls it. He might 'ave been hurt!'

'But,' Bimmie cried, 'it was Lee Dockerty, who got knocked out.'

'Blow Lee Dockerty!' cried the indignant Baker. 'I'm talking about Matilda. If he hadn't been extra sharp that night, he could have got hisself injured. He's like a son to me, that animal, after all we been through together. And then sendin' in a ringer to do him a mischief.'

'But we didn't send him,' Bimmie protested, 'he went—I mean, he came—I mean, it was just like Duke Pockhurst wrote in his column all about it. Matilda's famous now, can't you see? We're gonna clean up.'

'Fymous, is he?' Baker said. 'Too bloody fymous! Every yokel in the County wants to tyke a crack at 'im after me exhibition. And just last night we missed another pro having a go. Lucky Matilda zapped him first crack out of the box. He admitted it, when he recovered consciousness half an hour later.'

Patrick Aloysius was looking thoughtful. 'From my perusal of the article, Matilda handled Dockerty like a baby and from what I seen of him that day with you, he could take on a dozen of them hayseeds and a couple of pros as well in a night. What have you got to be afraid of?'

'Afraid?' Baker echoed. 'Oo's afraid? We're a circus act and not a bloomin' set-up who every sod that wants to get hisself some publicity can tyke a punch at. Matilda, he's the greatest. He can take care of hisself, but billin' him as middleweight champion of the world . . .'

'But he is!' Bimmie cried. 'Mr Pockhurst says he is. Something to do about no boxing law in Mississippi and he got a knockout according to regulations.'

Patrick Aloysius was tugging at his pendulous lower lip and eyeing the Cockney trainer with what almost amounted to suspicion. 'Well then, what's biting you, Billy?' he asked.

'What business is it of yours?' Baker growled. 'I've got a valerble property here. A year or two more and I can retire, buy me a pub back home and neither of us need work any more. I've put many years of me life into making Matilda what he is today.'

'But that's just it,' Bimmie interrupted, 'you've done it. Can't you see? You've made him the middleweight champion of the world. Mr Pockhurst got under Pinky Schwab's skin and he's going crazy. He don't know what to do because when Mr Pockhurst says something in his column about anything, that's how it is, and he was there. Ain't that so, Patrick Aloysius?'

Ahearn said, 'Pinky did seem to be a bit put out. It isn't every day you get your world's champion stiffened out of town by a kangaroo. Is that really what happened?'

'Sure it's what 'appened,' Baker replied indignantly. 'I knew right away, soon as he puts his hands up he wasn't no amateur. But to Matilda he was just another bloke who could give him a bit of a work-out and he was enjoying hisself until the bloke got nasty and Matilda zapped 'im. But supposing Dockerty had got through with that sneak punch?'

'But Mr Pockhurst said he didn't and Matilda's the champ now. You know what a world champ can make today—especially if he's a kanagaroo? A million dollars! Everybody says so. Me and you got a goldmine. Remember now when you first come in to see me, looking for a job and said about how what you wanted to do was to go back and buy your granpaw's pub in England? You could buy up a whole chain of pubs with all this publicity Mr Pockhurst is writing. Anyway, you don't have to go on with the ack any more. You got to resign from Mr Matson. We'll get some real fights for Matilda and after he's knocked out maybe half a dozen or a dozen, Pinky gotta get back in the ring with us for a million bucks.'

'A million dollars?' Baker said, and it was evident that he was startled. 'What are you blokes doing, pullin' me leg? Anyway, 'oo says I got ter resign?'

'I do,' said Bimmie. 'I'm Matilda's manager now. My contract says I got fifty-one percent, so what I say goes. I got you a second's licence so you can go in the ring with Matilda and look after him when he fights.'

A most curious expression came over The Bermondsey Kid's battered countenance, almost as though he were trying not to laugh. He said, 'So you're his manager now, are you? Well, good luck to yer, and what's him here got ter do with it?'

'He's just helping us out, giving advice, see? Because he's got a stable of fighters and knows a lot about the game. He's a friend of mine.'

'Friend, my royal tochas!' said Patrick Aloysius Ahearn

81

succinctly. 'You've got the greatest puncher there I ever saw. I've been trying to buy in. I've offered Bimmie here twenty-five grand for ten percent. What about selling me a piece of your share? With the right kind of build-up, Matilda would be the biggest thing that ever happened to the fight game.'

The Bermondsey Kid now looked from one to the other of the two men and said, 'Blimey, if I don't believe you two gents are serious!'

'You can bet all the tea in Formosa we are,' said Patrick Aloysius. 'Anyone, man or beast, who can put Lee Dockerty into the deep freeze in two rounds, could clean up the whole middleweight division in six months. We take in a quarter of a million at the gate building him up; another hundred grand selling tickets to his workouts, and then the gravy.'

For a moment still, Billy Baker glanced from one to the other and then dropped his voice and said, 'I think I'd better have a word with you two gentlemen in private.'

Without knowing exactly why, Patrick Aloysius had been waiting for something like this. He nodded his head in the direction of the bathroom.

'What's the matter with here?' Bimmie said.

'The walls of this here hostelry appear to be made of rice paper,' Patrick Aloysius said. 'Besides, if it's something important, the can is the spot. The one and only Tex Rickard in his heyday never concluded any important bit of match-making or boxing deal except from the throne. It was the only place where the press couldn't follow.'

Solemnly the three men filed into the bathroom and for the next twenty minutes the only sound that emerged therefrom was that of running water, as Patrick Aloysius had turned on both taps to cover their voices.

When they came out there was a material and psychological difference in their relationship, sealed by a brief but clearly binding document written on the back of an envelope and signed by all three. The fame and future earnings of Matilda, designated as middleweight champion of the world, were to be shared one third each by Solomon Bimstein, William Baker and Patrick Aloysius Ahearn unequivocally and covering every aspect of such earnings from whatever source including, as Ahearn had once seen written into a moving picture contract, the 'not yet invented'.

And as if this sudden alteration in the status and the inclusion of Patrick Aloysius Ahearn were not mystery enough, no money had changed hands.

It was about a month later that the messenger from the ninth-floor receptionist shambled into Duke Parkhurst's private office in the sports department, where he was engaged in composing his column for the next day's paper, and laid a receptionist's slip on his desk.

It had been filled in so that it read:

MR/MRS/MISS Solomon Bimstein (Bimmie)
WISHES TO SEE: Mr Duke Parkhurst
STATE NATURE OF BUSINESS; Important.

Parkhurst looked up from his typewriter and said, 'Who the hell is Solomon Bimstein?'

The messenger mumbled, 'I dunno,' and then added, 'he's Jewish.'

'Go on!' said Parkhurst. 'Important business, eh? Important for Mr Bimstein. Oh well, send him in.' When writing his column it was not par for the course unless he was disturbed at least four times.

He sat back at his machine. He was not expecting that he was going to like Mr Bimstein; he was not disappointed at first appraisal.

Bimmie was wearing the first of a series of garments approximating his idea of the well-dressed Broadway celebrity, a brown checked affair that did not suit his sallow skin or wiry blond hair that now arose at the sight of this great and commanding man sitting at the instrument of his greatness and power. Upon entering the office he stood nervously on the threshold revolving in his fingers a dreadful, mustard-colored cloth hat with a green feather in its band and a brim so narrow it was hardly more than an edge. The effect was not good.

Bimmie said, with what he hoped was an ingratiating smile, 'Mr Pockhurst, I'm Bimmie.'

Parkhurst examined the scrawl on the receptionist's slip

and said, 'So I gather. Well, Mr Bimstein—or Bimmie, whichever you prefer—what can I do for you?'

Bimmie was trying not to be overwhelmed by the trappings of the office, the deep pile carpet, the great blown-up photographs of thrilling moments in various sports, the air conditioning, the three telephones on the desk, the IBM electric typewriter, and the total grandeur of 'Mr Pockhurst' himself.

Bimmie's eyes strayed to the sheet of copy paper with two carbons emerging from Parkhurst's machine and he asked eagerly, 'Are you writing about Matilda again—'

Parkhurst glanced at the half completed column which was not about Matilda, it was raking the Inter-Collegiate Athletic Association over the coals, and said, 'No, not today, though I may again another time. Why? What's on your mind?'

Bimmie was too full of his great idea which he was going to present, over the objections of Patrick Aloysius, to hold it in any longer and so it all came out in one blurt: 'Mr Pockhurst, I would like if you'll accept it and enjoy it afterwards in good health, that you should have ten percent of Matilda. You wouldn't have to do nothing except go on writing like you have been about Matilda being the world's middleweight champ, and we wouldn't have to put anything on paper, Mr Pockhurst, because you can trust me. If I say you get ten percent, it's like you have got a bank statement. Once a month you could have it in cash, whatever it is we're making, which if you could go on writing like you are, will be plenty, so nobody would lose by this. Like I said to Mr Ahearn, what's ten percent to us, for showing our graditude, when we can pick up four or five grand for a fight? If you tried to buy what Mr Pockhurst writes, it would cost you a fortune. And all we would wanna ask is that maybe once a week you would put in your column something about Matilda, or maybe if you feel like writing more, you could do so.' He ran down for a moment.

Parkhurst shook his head with somewhat the same motion of a boxer who has been stunned and is trying to clear the haze. He said, 'Let me get this straight, Mr Bimstein. Are you suggesting that you will pay me ten percent of the earnings of Matilda in exchange for my writing a column about this animal at least once a week, with permission to do more if so inclined and that then you will hand this over regularly in cash?'

'You got it, Mr Pockhurst,' said Bimmie, 'you got it the first time. And if you want, we could shake hands on it,' and he held out a hand whose fingernails did not quite match the gala of his attire. He never knew how near death he was at that moment.

For Parkhurst was beginning to breathe heavily through his nose and an inhuman glare had replaced his contemplative inspection of his visitor. He said, in a tight voice, 'Bimmie, look behind you. Do you see that crack in the panel of that glass door?'

Bimmie looked.

'That happens to be a wire-reinforced fire door and the glass is half an inch thick. That crack was made by the head of the last son-of-a-bitch who tried to bribe me, when I kicked his ass out of here. You've got just three seconds to get through it on your own feet before you do it on the fly.' And he arose from his chair, huge, menacing and righteously indignant. The inviolability of the sportswriter, editor, columnist and his own department from any kind of venality was a fetish with him. Only writers whose hands were a hundred percent clean and who wore no man's collar could serve the public and the readers of the *Mercury* as they deserved to be served.

Bimmie turned back to face the giant and became as paralysed and unable to move as a budgerigar transfixed by a black mamba. The usual cliché of being rooted to the spot was hardly adequate, nor did 'turned to stone' describe it fully. It was simply that life had departed from his body. His bulging eyes gave him the air of some absurd species of parrot. He had not the slightest doubt that he was going head first through the wire-mesh, half-inch, glass fire door.

However, with his limbs immobilized and unable to flee from rapidly approaching total annihilation, his vocal chords yet had a few vibrations left and in terror he squeaked, 'Mr Pockhurst, Mr Pockhurst, don't do anything! It wasn't no bribe, it was just gradpitude.' And then he said probably the one thing which by its sheer inanity threw Parkhurst off his stride, 'You don't know what you done for me, Mr Pockhurst, you've got me my girl. Mr Lebensraum said he wouldn't mind if I come in to call, if I'm out by eleven and the light stays on in the parlor.'

This way-out from left field introduction of a love story and someone named Lebensraum, coupled with the abject terror into which he had thrown the young man, suddenly drained all of the choler out of Parkhurst. There was even something oddly touching about the boy and when he had cried that he had not meant it as a bribe, Parkhurst had realized that he was telling the truth. Instead of letting fly with his number fifteen shoe, he sat down on the edge of his desk and said, 'Who the hell are you, anyway?'

The return to life of the young man was signalized by the rapid rise and descent of his Adam's apple and once more the turning of the hat in his fingers. Parkhurst reached over, took the dreadful thing from him and laid it on the desk behind him.

Bimmie's voice was almost normal as he replied, 'I'm Matilda's manager.'

Parkhurst said, 'Since when? The last I saw of him he was being handled by a little Limey ex-pug from Bermondsey named Baker. When did you muscle in?'

'I didn't muscle,' Bimmie said. 'Baker come to me when he was down and out and he and his kanagaroo had had nothing to eat for four days, and I staked him. See, I was a booker of theatrical acks, "Sol Bimstein's Attractions— Book with Bimmie." It was me booked them into Matson's Monster Carnival where you seen him knock out Lee Dockerty.' And thereupon, without drawing breath he launched into the whole story.

'I see,' said Parkhurst, when Bimmie at last had run down, 'and when Matilda got himself some publicity, you moved in.'

'No sir,' said Bimmie, 'I had a contrack with Baker but see, that's all changed now. We got a partnership three ways; me, Patrick Aloysius Ahearn and Mr Baker.'

Surprise showed on Parkhurst's face. He said, 'So Patrick Aloysius has gone in with you. I wonder why?'

Bimmie said, 'Because he thinks Matilda's the greatest. He never stopped trying to buy a piece of him off me ever since the first time he saw him.' But he did not think it yet necessary to apprise Parkhurst as to Ahearn's function in the triumvirate and the reason for the partnership.

'And so what do you figure on doing?' Parkhurst asked. 'Okay, you can sit down.'

Bimmie took a chair gingerly. Parkhurst returned to his own swivel and leaned backwards, his hands clasped behind his head.

'Well, Mr Pockhurst,' Bimmie said, 'See, I thought I would ask your advice. See, we got a great property here on account of what you wrote and Patrick Aloysius—that's Mr Ahearn—says that we will knock out any middleweight on two legs or four, with one hand behind his back. We already knocked out some jig named Carter in thirty-seven seconds and another boy by the name of Joe Cisco in a minute and a half. But it ain't gonna be easy to get fights for Matilda, unless like now in your column you keep on saying he's the middleweight champion of the world. But I swear, Mr Pockhurst, that wasn't what I said the ten percent for.'

Parkhurst nodded absent-mindedly. The absurdity that something which had started half as a rib and a needle designed to irritate several unsavory characters should suddenly take on substance! It had provided him with great running copy; he had published the life story of Matilda, delved into the origins and characteristics of the species and enjoyed a large and growing fan mail on the subject. But now for the first time all sorts of different ideas, angles and an utterly mad notion were coursing through his head—an inspiration that might lead to only heaven could tell what triumphs of screwball journalism as well as promotion.

The thoughts were so dizzying that he compelled himself to go back for an explanation of the crazy thing Bimmie had said when he thought he was faced with instant death, namely what someone named Lebensraum had to do with it all and why the lights had to stay on in the parlor.

He asked, 'Tell me, Bimmie, where does this party by the name of Lebensraum figure in this picture?'

'Mr Lebensraum?' Bimmie repeated. 'He's the old man—

the father—of my girl Hannah. See, when I was booking acks things weren't going so good and Mr Lebensraum says to his daughter who comes from a very good fambly, how she shouldn't be going around with a bum who when they go out to dinner, she has to pick up the check. What kind of a life is this gonna be? So she got to sneak out and meet me downtown. But after Mr Lebensraum, who follows your column like it's the Torah—he would miss his breakfast rather than what you're writing, read what you wrote about Matilda and then found out I was his manager and had a business partnership and a manager's licence...'

'How did you get the manager's licence?'

'Mr Ahearn got it for me.'

'From Colonel Wildman?'

'Yessir.'

Parkhurst's thoughts notched up another thousand revs and, as he fell silent, Bimmie continued.

'So Mr Lebensraum decided maybe I wasn't such a schmo and could be someday I'll have my name in your column too, and so I would be allowed to call regular on Hannah but no monkey business.'

Parkhurst's ideas were reaching so far out into the universe of New York's sporting life that he wanted another moment's respite to organize them and so he asked, 'Is Hannah a nice girl?'

'Oh gee, Mr Pockhurst, she's terrific—a doll! Boy, she's got them eyes like they describe in the Bible.'

'From the Song of Solomon?' Parkhurst enquired.

'Yeah. Well, sort of.'

'No, I mean she is really a nice girl?' Parkhurst went on. 'Is she good and...'

'Mr Pockhurst,' Bimmie assured him earnestly, 'you should only hear how she goes on about what's honest and I should never do nothing which ain't right, because that's the one thing up with which she would not put. They don't make them like that any more today, Mr Pockhurst. She comes from a great fambly—her mother was a Woolf.'

There was a good deal more concerning Hannah's merits which Parkhurst did not hear for once more that mad idea was visiting his brain and he hardly dared contemplate it. On the face of it, of course, it was impossible, and yet...

He cut Bimmie off in mid description of Hannah's ankles which were not thick but slender like a race horse's, and she did not have one of them big cans on her either....

He leaned forward, pointing a finger at Bimmie and said, 'Look here, Bimmie, we'll forget about the ten percent and just pretend that you never mentioned it. The trouble with you and your pals is that you don't think big enough, you're playing for peanuts. I like what your girl Hannah has to say about being honest because that's the way we try to operate here. I don't know how it was in the theatrical booking business, but the boxing racket is a sewer. A phenomenon like Matilda probably happens just once. If you keep your nose clean perhaps I'll go along with you for the jackpot.'

Bimmie's hair rose another notch and his blue eyes threatened to pop from his pale face. He said, 'You mean like a silent partner?'

'Not so silent,' Parkhurst said, and motioned towards his typewriter.

Bimmie was filled with a sudden rush of virtue. Thoughts of Hannah and her trust in him that he would not do anything that was wrong ennobled his spirit and now on the same basis, this great man was offering him some kind of partnership; referring to four or five thousand clams as peanuts, conjuring up visions of endless dollars pouring forth as from a fruit machine. It was on the tip of his tongue to tell Parkhurst what had transpired in the john with Billy Baker and Ahearn that day in Kiowa, so that his new-found ally would understand the situation completely, when he heard himself repeat the word, 'Jackpot?' the query had slipped out almost without his realizing it, even while he was making up his mind to live up to the high standards of Mr Pockhurst.

The columnist grinned at him and said, 'It could be a million dollars for your end.'

Those two words, 'million dollars' rang through Bimmie's head with the clang of a golden gong and momentarily suspended the rush of rectitude that had seized him. To be sure he had used that same sum himself but it had only been to con Billy Baker and was no more than normal

Broadwayese exaggeration. On the Main Stem if you talked about a million you meant maybe fifty or a hundred grand; when you referred to an act as being colossal you meant that it might get by. But that sum dropping from the lips of the Duke himself? Forgotten was what he had meant to confess as his own whirling brain once again took off on flights of fancy—gifts showered upon Hannah, cigars brought in from Havana by way of Britain were offered to Mr Lebensraum, real estate agents consulted about property in Miami Beach.

'If I ever hear about you doing anything dishonest,' Parkhurst was continuing, 'I'll flay you and nail your skin to the top of the Empire State Building and what's more, you'll never get into Mr Lebensraum's parlor again!'

Bimmie had not even heard the threat. His auditory nerves had closed down after the mention of the amount of the jackpot in the offing. He murmured, 'A million bucks!'

Parkhurst emphasized, 'I said *could* be a million dollars, but a hell of a lot's got to happen first and maybe it never will. But that would be your end eventually, if we could get Dockerty back in the ring again, plus whatever you can make on the way up—maybe another half million when you get close to the top. But you've got to have some guts, too, if my idea's going to come off. Do you know who owns Lee Dockerty?'

The other half of Bimmie's brain that was not computing the interest on a million and a half dollars came into operation again, and he replied, 'Sure, Pinky Schwab.'

Parkhurst shook his head. 'You've got a lot to learn, Bimmie. Ask Patrick Aloysius. Dockerty's contract is owned by a man who around Broadway is only referred to as Uncle Nono. He's boss of the Eastern U.S.A. branch of the Marching and Chowder Association of the Mafia. That means the real thing. I understand he blew a gasket when he read my pieces about Matilda flattening Dockerty. He's going to be even sorer when he catches onto what I've got in my mind. He and his mob don't dare do anything to me; he might to you.'

That did not penetrate either. Mr Lebensraum at the wedding was saying, 'Meet my son-in-law, Bimmie, manager of the world's middleweight champion. He's a millionaire.

You got any more of those cigars on you, Bimmie?'

Parkhurst was continuing, 'Could be I'm wrong and I want to talk to my boss about it first, but on the other hand maybe a chance like this comes once in a lifetime. God might just be fed up with these flesh peddlers and merchants of injury and Matilda the instrument of His vengeance.'

Bimmie asked, 'What are you going to do, Mr Pockhurst?'

Parkhurst grinned mirthlessly, his idea had taken a firm hold by now and it was frightening even him. He replied, 'Arrange a little press conference. Watch the papers. If it comes off, keep on like you're doing with that animal, but let Patrick Aloysius pick his opponents. He knows more about the fight game in his little finger than you'll ever learn in a lifetime. You handle the business end. You're smart enough, or you wouldn't have got this far. But remember, you'll more than likely find yourself in some jams when Uncle Nono and his boys begin to operate. Ahearn will understand that. It's up to you to get yourselves out of them without crawling to me. Afterwards if you're still around, I'll print the story. If I get away with this I'll put in a claim for a Pulitzer Prize.'

Parkhurst was only half kidding with the last statement, but being human like Bimmie, he too was having certain dreams of grandeur. It was not only that a phenomenon such as Matilda occurred only once in that lifetime, one had to know what to do with it when it did happen.

And to Bimmie the prospect that opened up was so dizzying that everything else quite went out of his mind and he could only say, 'Gee, Mr Pockhurst, I wouldn't know how to say thank you. I'll do like you say; I'll do everything like you say. You wouldn't be sorry, Mr Pockhurst.'

'Okay, then,' Parkhurst said, 'beat it. And whatever you do, keep your mouth shut.'

'Numb's the word, Mr Pockhurst, and thanks again.' Bimmie went out of the door but stuck his head back inside once more to say, 'We couldn't give you just five percent, Mr Pockhurst, could we?' And he was just in time to withdraw head and mustard-colored hat as with a thunderous crash the Manhattan telephone directory arrived at the spot where they had been.

# X

FOLLOWING A PERIOD in the doldrums, boxing in New York had been brought to life again by a live-wire businessman, Walter Mason, who had been one of the large stock holders and vice-president before he took over the presidency and management of the Metropolitan Arena and the Jericho Stadium. The Arena, built out over the railroad tracks at 66th Street on the Hudson River, seated twenty-five thousand for fights. The capacity of the Jericho Stadium on Long Island was over a hundred thousand.

His involvement, as Parkhurst suspected, was inevitable. Three classifications were the most important in boxing: the lightweight, the middleweight and the heavyweight. To have allowed the value of any of these to become tarnished would have been to have downgraded a commercial asset.

The continued straight-faced references in the sports pages of the *Mercury* not only in 'The Duke's Deadline' but also in the stories by Murray O'Farrell, Parkhurst's scrappy little boxing writer, to Matilda as the middleweight champion of the world eventually were bound to call for some sort of rebuttal. This was particularly so because the Duke's campaign was having a secondary effect which was even more alarming to Mason and the corporation backing his promotions. Due to Parkhurst's readership, reputation and wide syndication a large section of the sports public was becoming brain-washed and accepted someone by the name of Matilda, if not a kangaroo, as the legitimately crowned middleweight champion of the world. Any time Dockerty got into the columns of the *Mercury* he had ex-champion appended to his name.

92

Mason was finally goaded into producing not only Pinky Schwab but also Lee Dockerty for a press conference attended by all the boxing correspondents, even though they had no intention of writing about it. The Matilda affair was considered Parkhurst's joke and hence not to be touched with the symbolic ten-foot pole.

Nevertheless, when Mason announced that on a certain afternoon Pinky and Lee Dockerty would be present in his office for questioning, they turned up in force just to see the fun. Naturally they expected Parkhurst to be there and were looking forward to the prospect of his needling Pinky, Lee and Mason himself. However, Parkhurst was too subtle for them. He stayed away and sent only Murray O'Farrell. To have attended would have been something like the Allies recognizing East Germany, acknowledging that Lee and Pinky had a case and still held the title.

Mason, who was inclined to take himself seriously, was a burly ex-football player and Yale graduate who had made his way upwards in the world of business and corporation management. He was a heavy set, rather sententious man inclined to ignore the lighter and more scandalous side of the business and an occasional target for leg-pulling on the part of the sportswriters. But withal, he was intelligent enough to hire a good matchmaker, put on fights that kept the Arena in the black and could recognize a drawing card that would fill his gigantic Long Island Stadium. He had promoted two of Dockerty's fights and both had been profitable. In his middle forties he was a young-looking, bouncy and ebullient executive.

Present, besides Mason, Schwab, Dockerty and O'Farrell, were Cassius Jones of *The Times,* Tom Petrie of *The Post,* Jack Horne of the *Daily News* and Gill Hobart of Long Island's *Newsday.*

Summer having burgeoned in the land, Pinky had laid aside his moth-eaten astrakhan and Lee had been persuaded to climb out of his jeans and greasy leather jacket and come respectably clad in trousers and an open-necked shirt.

Mason brought the meeting to order by rapping on his desk with a miniature gold football on a pedestal, verification of his one-time All-America selection, and began his address:

'Gentlemen, I have asked you here today in the interest of an injustice which I feel has been done to our friends Pinky Schwab and Lee Dockerty in the course of the publication of some articles which I consider crude, unfunny and in bad taste. Aspersions have been cast upon the validity of the title of Dockerty. It would not be necessary to dignify these remarks by acknowledging them except for the fact that they have appeared in a newspaper of so large and influential circulation that a number of people have come to give credence to this nonsense and our friends here, Lee and Pinky, have been held up to ridicule and scorn.

'Now Lee, here,' Mason continued, nodding his distinguished head in the direction of the fighter, 'is a good boy. He has worked hard to achieve the top rung of the ladder of fistiana. He has been a good champion and . . .' At this point he chose to overlook the soft raspberry blown by Tom Petrie of *The Post* who had written his exposé of how Lee Dockerty's build-up had been achieved, naming the fighters who had taken a dive for him. Mason continued, 'And I think, therefore, it is only fair to let him state his case and tell what happened in his own words and I am counting on you fellows to give him a break and print it like it was.'

Cassius Jones, the dean of the boxing writers, said, 'I wouldn't count too much on that, Walter, why should I bother to deny something I've never stated?'

Tom Petrie said, 'We might, if it's funny enough. Okay, Lee, go on then, give! What did this animal hit you with? Or did you make the whole thing up as a publicity stunt?'

It was a bad beginning and Pinky at once moved in with, 'Well, now, see, it wasn't at all like Duke . . .' when Gill Hobart cut him off with, 'Oh, shut up, Pinky, and let your boy talk for once. He's the one it's supposed to have happened to.'

Dockerty struggled for a moment in a desperate attempt to retain the speech he had been coached to make, found that it had completely escaped from his memory and mumbled, 'Uh . . . uh . . . You see I was down there back home—I got a little business there to look after with a coupl'a fellas, and there's this Fair like, so we went and had a coupl'a drinks . . .'

'Cokes, you know,' cut in Pinky quickly. 'Lee don't never touch no alcohol.'

Jack Horne of the *News* said, 'Parkhurst never said you were drunk in his piece.'

Gill Hobart added, 'So what did you do, tie a can on and think it was funny to slug this dumb brute?'

'Nah!' said Lee. 'Like Pinky said, we had a coupl'a cokes . . .'

'And you're allergic to caffeine,' put in Petrie.

Dockerty did not know what that meant, so he went on, '. . . So there's this thing about a boxing kangaroo so we went in, maybe for laughs. So then when they offered five hundred bucks for anyone to stay in there for a coupl'a rounds, Chick—one of my buddies—says, "Why don't you . . . uh . . . uh . . . go in there and . . . uh . . . uh . . ."' Here Dockerty trailed off as a genuine memory assailed him and he realized that a description of what the five hundred dollars was to be used for might not go down very well at this interview.

Cassius Jones of *The Times* produced the back of an envelope and a pen from his pocket and made a note. He said, 'Five hundred dollars—so there was a purse. That rather made it somewhat official, didn't it?'

Mason was beginning to tap nervous fingers on his desk. He said, 'Now wait! Let him finish . . .' The interview was not going at all the way he expected.

Pinky Schwab also was beginning to be flustered. He said, 'Maybe I ought to tell it. Lee ain't much at making speeches.'

Murray O'Farrell reached back some thirty years or more for a line that had been made popular over the radio by a German comic in those days, but not so far that it did not register with Pinky. He asked, 'Vas you dere, Scharley?'

Pinky turned on him, 'Listen, you little bastard . . . !'

Mason rapped the football on the desk top again and said, 'Now, now, we want everything to be friendly here. What we're all doing is trying to help. Go ahead, Lee, tell it in your own words.'

Dockerty said, 'Uh . . . he hit me when I wasn't looking.'

Gill Hobart giggled and asked, 'Where were you looking, Lee? Was there a broad at the ringside?'

Dockerty said, 'I mean, he hit me low. I was fouled.'

Tom Petrie enquired, 'What round did this happen?'

'Inna second round.'

'What were you doing in the first round?' Jack Horne enquired.

Pinky was in fast with, 'It always takes Lee a round to warm up. See, he didn't wanna hurt this animal. Lee loves animals. He was just kidding around.'

'But he was on the floor at the end of the first round,' Hobart commented.

'I slipped on a wet spot,' Dockerty said. 'The referee didn't count...'

Cassius Jones made another note on the back of his envelope, 'Oh,' he said, 'so there was a referee. Was there a referee, Lee?'

'Uh... well... there was a fella there like, who acted like a referee.'

Gill Hobart suddenly asked, 'Was there a weigh-in?'

'Uh... well... Yeah, they said you had to be middle-weight...'

'Lee was overweight,' Pinky put in quickly, 'he was 165 pounds. When you been drinking a lot of beer...'

'Coca-Cola, Pinky,' Petrie corrected solemnly.

Gill Hobart said, 'Lee never weighed more than 160 pounds in his life.'

Horne asked, 'What did the kangaroo come in at?'

'159 even,' Dockerty replied, not catching the signals Pinky was wafting at him.

Pinky then added, 'But the scales was gimmicked.'

Cassius, his pencil poised, asked, 'Did they ask you to sign any kind of a paper, Lee.'

Dockerty scratched his head and then replied, 'Uh... Yeah, that's right. I had to sign that if I got hurt I wouldn't make no trouble for anyone. I couldn't sue, see? They said they had to do that with everybody in case of accident. And if I stayed two rounds, I got paid.'

Cassius Jones nodded, scribbled again and said, 'So one might say there was a purse, a referee, a weigh-in and a contract. Sounds pretty regulation to me.'

Mason twisted heavily in his chair, 'Now see here,' he

cried irritably, a good deal of the velvet departing from his voice, 'who are you fellows for? Lee here, or that son-of-a-bitch Parkhurst?'

With the dignity befitting his years and the paper for which he reported, Cassius Jones replied, 'What we're for is the truth.' He then continued, 'Now about this foul that you claim, did the referee see it, or warn your—ahem—opponent?'

'Nah,' Lee replied and looked uncomfortable.

Cassius Jones said, 'According to Parkhurst you tried to clip Matilda on the break. Maybe that's the foul you're referring to?'

The tall, thin, white-haired Timesman exuded the aura of a gentleman. Lee was a little afraid of him and hence grew more confused, 'Yeah, yeah,' he said, 'the son-of-a-bitch was hanging onto me and licking my puss. He stank like a goat so I thought I'd clip him one. I got mad.'

'And what happened then?'

Dockerty was mumbling again, 'I dunno. It's like I suddenly felt mum all over. Maybe he poisoned me with them licks. It was like I was in a transom.'

'But he hit you,' Horne said.

'Yeah, I guess so.'

'How long were you out for?'

'I dunno,' Lee Dockerty confessed, at which point Pinky Schwab blew up, his eyes crossing, his face contorted with rage, saliva drooling from his mouth.

'It's a lot of goddam lies! The whole thing is a pack of lies. You're getting my boy all twisted up like with a cross examination. This ain't no court. Duke Parkhurst is nothing but a rotten liar! Like Lee told it to me, the whole thing is a gag and clowning around, and Lee goes in there to do them a favor on account of who he is, and goes along with them by pretending he's kayo'd. The whole thing is an act they cooked up out of the kindness of Lee's heart and now this lousy bastard writes a lot of lies which ain't true about my boy, and if I ever see him to his face...'

'... You'd better duck, if you call him a liar,' Cassius Jones finished for him. 'Cool it, Pinky, when a man with Parkhurst's reputation lays something on the line that might injure someone or attract a libel suit, he's apt to be pretty sure of his facts. You could sue, you know, if you thought he had made any misleading or untrue statements, though I gather from his piece that he had plenty of witnesses.'

Pinky made an attempt to swallow his choler which popped his eyes still further. He said, 'I wouldn't dignify the man by suing.'

'That's right, Pinky,' Mason soothed, 'it's not worth the trouble.'

Tom Petrie asked, 'Then what do you want of us? What are we all here for?'

Mason said smoothly, 'You fellows ought to write that Lee Dockerty is still middleweight champion of the world.'

Horne snickered and said, 'We never said he wasn't. He's just not middleweight champion to Duke Parkhurst and Murray O'Farrell here, and the readers of their lousy sheet,' and then he added, '... two million of them.'

Petrie added, 'He's syndicated. Twenty million of them.'

Pinky Schwab was warming up to another explosion, 'Listen!' he squealed, 'when a man wins the middleweight championship of the world, he's got to be champion for everybody, don't he?'

Gill Hobart said, 'Maybe not, if he's just been knocked on his ass by a trained animal. What the hell kind of a champion is that?'

Little Murray O'Farrell was heard from again. He said, 'For us he ain't champion unless he goes out and proves it again.'

Pinky's voice rose a pitch higher, 'What? Against a stinking kangaroo that ain't even human? What do you expect my boy to do, fight lions and tigers and elephants? Have you guys all gone off your nuts?'

Horne said, 'The world takes in a lot of territory, Pinky. If anybody or anything can stand up to Lee wearing a pair of gloves and flatten him, he's got to be heard from, doesn't he?'

All the sportswriters were grinning happily. A successful needling of anyone, but in particular Pinky Schwab was affording them joy ineffable and Cassius Jones asked solemnly, 'Would you fight him again, Pinky?'

A pressure gauge attached to Pinky would have indicated

that the danger mark was rapidly being approached. 'Listen!' he shouted, 'Lee don't draw no lines. We'll fight anybody on two legs who...'

Hobart said, 'Well, according to Parkhurst, the 'roo was on two legs when he clobbered your boy.'

Tom Petrie, the afternoon paper columnist who was known as the best ribber of the bunch started off on a new tack. He wanted to preserve this pigeon a little longer before he blew his stack completely. 'Now, wait a minute, fellows,' he said. 'Let's all be reasonable about this and give Lee and Pinky a break. Let's look at it from another angle. This sort of thing happens all the time, doesn't it? A champ goes out of town for a series of exhibition matches or takes on two or three local boys at over the weight, has an off night and gets knocked down a couple of times, or maybe runs into one of those home-town decisions. So what happens? You've got a great build-up for a return match, the champ stiffens the home-town boy and nobody's hurt. Isn't that about the straight of it, Pinky?'

The fat manager, momentarily seduced by Petrie's dovelike cooing and unexpected sympathy said, 'Yeah, yeah, sure! I guess so.'

It was now Mason who was beginning to hoist distress and warning flags, and trying to catch Pinky's eye. He could see the trap into which Pinky was being steered but it was too late.

'Well, then,' Patrie said, 'you saw where Matilda's manager said they'd give Lee a return bout, maybe in a year. All Lee has to do is to go in there, knock him out and...'

The jaws of the snare clanged shut. Pinky Schwab hit the ceiling. '*They'll* give us a return bout! *They* will! Who the hell are *they?* Who did they ever lick? Let them go out and get themselves a reputation. My boy fought his way up from the amateurs. That thing ain't in no boxing record book. It's a circus act that lands a lucky punch and now he's going to give *us* a return match!'

Murray O'Farrell, who had been primed by Parkhurst for many eventualities during the interview but had not expected this piece of manna from heaven, said, 'You meant if they went out and got themselves a reputation like you say,

Pinky, got a build-up and a record of kayos of some good boys, you'd take 'em on?'

Petrie turned the screw of the vice. He said, 'Or okay, then, Pinky, put it the other way round. Supposing we wrote that Lee is still champ, but that this Matilda looks like the best prospect around, if he could get maybe half a dozen or so of the better middleweights in the business, you'd have to fight him, wouldn't you?'

Even Pinky saw the pit into which he had fallen by now and could only glare like a cornered animal, the breath of wrath and frustration whistling through his nose. He snarled finally, 'I said Lee wouldn't bar nobody. He'll fight anybody if there's enough dough in it.'

Walter Mason scraped his chair, coughed, fussed with the papers on his desk for a moment in those gestures that are used to terminate a conference. The outcome of the interview had been quite different from what he had envisioned. He had called it merely for the purpose of protecting his investment in Lee Dockerty. Due to the idiocy of Pinky Schwab, that investment had unexpectedly suddenly taken on an added value and all warning flags were struck. As a rational businessman it did not seem likely, but as a promoter he could not help but figure how much a bona fide return match between Lee and this Matilda thing could draw into his big outdoor Stadium. He said smoothly, 'Look, gentlemen, there's no need to hurry on this. We can go into all this later. If those fellows want to go out and build themselves up a reputation... Well, that's their business. In the meantime, gentlemen, I think you've all been very fair. I thank you and I hope you've got your story.'

'Oh brother,' said Murray O'Farrell, as they filed out, 'have we!'

# XI

THERE WAS NOT a line in the other papers the next day although the four boxing writers Cassius Jones, Petrie, Horne and Hobart pleaded with their editors to let them go to town on the story of the discomfiture of Pinky and Lee Dockerty. Their theory, good journalistic policy, was that it was too good to miss and if you can't lick 'em, join 'em and, if possible, even take the story away.

But the editors were still wary of what Parkhurst might be up to or what he had in mind and therefore missed the opportunity to get aboard what was showing signs of becoming a band wagon. They let the *Mercury* score another clean beat with the story by Murray O'Farrell, headlined, 'LEE DOCKERTY—MATILDA RETURN MATCH BUILDING. Ex-Champion Bars Nothing On Two Legs.'

O'Farrell's story began: 'In another startling and unprecedented turnabout yesterday in the lush office of Walter Mason's Metropolitan Arena, Lee Dockerty ex-middleweight champion of the world, through his manager Pincus (Pinky) Schwab, agreed to meet Matilda newly-crowned title-holder, in a return match after the latter had confirmed his victory over Dockerty by taking on and defeating some of the best middleweights in the country, who are already clamoring for a crack at the title.

'In a generous gesture that will earn him the plaudits of fistiana's fans, Schwab said that Lee would bar nobody, man nor beast; he will fight anyone.'

O'Farrell went on from there for another two columns expanding the story, rehashing Parkhurst's account of

Dockerty's knock-out at the hands of Matilda and estimating the gate that a rematch between the two might draw a year from then.

Parkhurst had purposely refrained from adding a copyright line to the story and so the Press Associations, whose business it was to disseminate news where they found it, and who have thus nothing to lose, picked it up and spread it over the country where it both entertained and stimulated a huge readership extending far beyond the confines of boxing fans. Arguments were initiated in the home as well as in bar rooms as to what would be the outcome of such a match and in an extraordinary way attracting an entirely new group which heretofore had not exhibited the slightest interest.

The piece had a particular effect upon two individual parties. In the south-west, when excitedly read by Bimmie, Ahearn, Baker, Matilda and Company, it acted as a green light. On the other hand, to an eminent businessman, socialite and notable New York sportsman it had the well-known effect of the red cape waved before an irritated bull.

On the morning that the story appeared two important meetings took place in two widely separated suites of offices, the first in the brokerage firm of Arbalest & Mitchum at No. 49A Broad Street, and the second one later on in the higher reaches of the Empire State Building at Fifth Avenue and 34th Street, at a corporation devoted to trade and commerce known as The Neapolitan Import and Export Company; President, Giovanni, (Gio) di Angeletti.

Arbalest & Mitchum did enough brokerage business so as not to attract any undue attention. It also provided an inner sanctum for Johnny Renato, a swarthy, sleek and dangerous little man who was the second-in-command and alter brain of the Capo who was overlord of Cosa Nostra's entire East Coast network, the shadowy, unseen, unknown figure referred to only as Uncle Nono.

Present at the meeting besides Renato were Pinky Schwab and Joe Marcanti, a Broadway character who ran a book, was chief collector for the 'numbers' game and at the same time held the boxing chair in the Mafia's eastern sport's division. He was a burly thug with the face of an aging choir boy. Present, too, were several Mafia elders—members of the fiscal and judiciary branch—whose function was to negotiate with the State, city or borough government when necessary, or to purchase a judge or a policeman or two.

For the moment Renato, at the head of a small board table around which they sat, was concealed behind the copy of the *Mercury* containing O'Farrell's story. When he put it down he did not speak immediately, but let the weighty silence grow heavier while he examined the polished fingernails of delicate and shapely hands. When he spoke finally it was in a voice that was deep and husky, slightly tinged with the accent of one who has worked his way up from the streets of one of the Italian ghettos. He said, 'Uncle Nono is not happy.'

Pinky, who was sweating so that he could feel his underwear sodden inside his clothing, began, 'Boss, I . . .'

Renato cut him off with, 'Shut up! I'm doing the talking.' He now tapped the sheets and said, 'You, Schwab and Marcanti, what the hell is this all about? Pinky, you must have been out of your mind to let Lee out of your sight to get into that trouble down south, and give Parkhurst the chance to write up all that crap. I thought you and Marcanti were supposed to be looking after him? That's what we pay you for, isn't it?'

Pinky pleaded, 'Boss, how could I tell? The kid said he wanted to go home for a couple of days. How can you figure a boy is going to get into trouble when he wants to go home where he was born? See, it's just a gag like I said there. Lee was trying to help these fellows out and . . .'

Renato turned a cold stare upon the fat man and said, 'Who are you trying to kid? Lee wasn't doing anything of the kind. We sent a man down there to check. Lee got drunk and made a horse's ass of himself. We were prepared to ride it out, Parkhurst couldn't go on writing about it forever, when you let yourself get suckered into that press conference where you make a further fool of us by agreeing to give this animal a return match. Is that right?'

The drops falling from Pinky's fat face were now forming a little pool at his feet. 'Aw, now, boss, I never said anything like that. See, those reporters are all a bunch of liars. They twist everything around that you say. Like now I never said nothing about any return match. I said they should go out and get themselves a reputation before talking about a return match.'

Renato tapped the paper again and said sarcastically, 'I can read,' and then turned on Marcanti and said, 'Where the hell were you? You should have been on top of this long ago. Who are these people?'

Marcanti said, 'Boss, it wasn't my fault. I figured Pinky knew, or at least he ought to have known, on account they all worked out of the same office—him, Patrick Aloysius Ahearn and this little Jew boy by the name of Solomon Bimstein—Bimmie—who used to book cheap acts into night clubs. This Paddy Aloysius Ahearn has had a stable of bums for years. Now he seems to have gone in with this Bimmie and an ex-fighter by the name of Baker, a Limey who trains this animal. The three of them are working together now. Pinky should have broke it up before it started, or given us a tip.'

The unhappy Schwab protested, 'How was I to know? I wanted to turn the little bastard out of the office long ago, except Paddy Aloysius said he wasn't doing no harm booking acts. The next thing I know, Parkhurst puts all this crap in the paper and . . .'

Renato shut him up with another look. He said, 'Can we talk to them? Make it worth their while?' It was always Mafia policy to buy its way out of trouble before proceeding to more drastic methods, which was the reason for the presence of the fiscal and judiciary branch.

'I don't know,' Marcanti replied, 'Paddy Aloysius has always been okay, but I don't know about this fellow Bimstein.'

One of the judiciary members said, 'We can keep Bimstein from getting a manager's licence anyway.'

Marcanti said, 'It's too late. Paddy's got him one already. That foul ball Wildman.' Then he continued eagerly, 'Listen, boss, if you want Bimstein hit, it wouldn't make a stink. Nobody's ever heard of him. He wouldn't even be missed.'

Renato said, 'I'll tell you when I want someone hit and how. Where are these fellows now—this Bimmie and the others?'

Marcanti replied, 'They're down south, trying to set up some build-up matches, like Pinky here said they should, the stupid son-of...'

'I never said they...' Pinky started to explain.

But Renato said, 'Shut up!' again, 'I'm talking to Joe.' He turned back to Marcanti. 'See that they don't get anywhere.'

Marcanti said, 'Okay, boss, I get it. How far would you want me to go?'

Renato reflected for a moment and then said, 'Like I said, Uncle Nono isn't very happy. If something were to happen to that Matilda so he would lay down, or quit, or get knocked out, or for some reason couldn't fight any more, Parkhurst wouldn't have anything to write about, would he? Down there in the south-west, where they haven't got too much law would be the idea.'

Marcanti said, 'What about these guys with him?'

Renato looked up at the ceiling and seemed to be talking only to himself now, as was his custom when terminal measures were in the air, 'Maybe a hit wouldn't be such a bad idea,' he mused, 'if they won't listen to reason. The way I see it, this Matilda should be persuaded to quit, maybe in so humiliating a manner that Parkhurst would drop him. If his people can't be persuaded to see it that way, well, they know the name of the game just as well as anybody else and will only have themselves to blame. Ahearn, at least, is supposed to be nobody's fool and ought to get the message. Have you got somebody in mind for the job?'

Marcanti did some quick reviewing. Two of the boys who had worked with him fixing fights which had contributed to Dockerty's rise to eminence were still with him and he could think of a third. He named them.

Renato now focussed the glitter of his gaze upon Pinky who was wishing he had never been born. The shenanigans of the fight game were one thing; the ordering up of murder *à la carte* another, in particular two men with whom he had been associated. Would it be his turn next?

Renato said, 'You're lucky. You're let off this time. I was against it, but Uncle Nono says you're a fool, but it wasn't your fault that Parkhurst's plane was forced down. If it hadn't been for that nobody ever would have heard of it. You're to get hold of Lee and keep him out of sight until this

105

blows over. Don't give Parkhurst a chance to mention him as ex-champion or anything else, until we get this thing killed off in one way or another. Keep your mouth shut and stay away from reporters and making yourself more of an ass than you have already.'

'Okay, boss.'

Renato was back on Marcanti again. He said, 'We wouldn't like anything to go wrong, Joe. Nothing Parkhurst could get a hold of for a write up in his column, for instance. Quiet and no trouble. And if there has to be a hit, it's got to look like some kind of a local double cross. These boys of yours . . .'

Marcanti said, 'They done some good jobs for us already. Like you said, Paddy Aloysius will get the message. It won't be the first time one of his boys will have lay down.'

Renato dismissed Pinky and Marcanti curtly, 'Okay, you two, that's all.' But to the elders he was more polite and said, 'Thank you for coming, gentlemen, but that's the way the boss wants it handled for the moment. I think this will clean it up. If there's any legal trouble, we'll let you know.'

When they had gone he waited for a moment, then put on his own hat and coat, went out, hailed a taxi and drove uptown.


The voice of Miss Benson, his secretary, came through on the squawk box on the desk of the President of the Neapolitan Import and Export Company, 'Mr di Angeletti, Mr Renato, the man from your brokerage office, is here.'

The President clicked the switch on the box and replied, 'Thank you, Miss Benson. Send him in.'

In his college days when Gio di Angeletti had been editor of *The Lampoon,* Harvard's comic magazine, he had been known for an agreeable sense of humor. In later years as a prominent merchant citizen this ability to laugh at himself, as well as the foibles of others, had become somewhat atrophied.

It was not so much the business of importation of olive oil, prosciutto, mortadella, salami and huge black Parmesan cheeses as big as cartwheels from Italy, which he had inherited from his father, that dampened di Angeletti's sense of the comic, but rather certain other responsibilities that went with the inheritance. Although Giovanni was born in America, the family for generations had been noted Sicilian businessmen, the business in which they engaged having been extortion, terrorism, smuggling, narcotics and murder.

Not until Giovanni had graduated from Harvard *cum laude,* soft-spoken, cultured and socially successful had he learned that his father's principal import into the United States had been the overlordship of the Eastern seaboard family of Cosa Nostra. Gio had been sent to Harvard deliberately to provide him with an unassailable cover when he should be called upon to take over the varied activities of the network.

At first his sense of humor saved him; the idea tickled him and when his father broke the news he roared with laughter. Later, when he had learned that there was no escape from his destiny except in a concrete envelope or ventilated with lead, floating face downwards outward bound with the tide from New York harbor, he laughed less. And when the full responsibilities of his high office and widespread organization settled upon his shoulders, he laughed not at all. He was no more entertained at having to order an execution than a duly appointed judge reading a death sentence from the bench.

Thanks to the schooling and early connections, he was extremely successful in keeping his two lives apart. The business was sufficiently large and lucrative to account for his wealth, his estate and breeding farm in Westchester, the town house on Upper Fifth Avenue, the villa and marina on Biscayne Bay in Miami. He was at that time a distinguished-looking gentleman in his middle fifties who kept fit by playing squash at the Harvard Club and exercising his own horses. He gave suitable parties attended by many of the best people, was amongst those present at first nights and big sports events, kept out of politics, contributed to charities, was generally a respected citizen and a part of the New York scene.

Yet loss of the ability to laugh was not the only occupational affliction from which Gio di Angeletti eventually suffered. The power he wielded, even that for which he had not asked, had inevitably rendered him vain. So successful had he been in keeping his two lives separate that via the shadowy figure of Uncle Nono, whose real identity had hardly been suspected, he was able to indulge his interest in sports in which in his Harvard days he had been a participant.

*Sub rosa,* Uncle Nono owned a professional football squad, a hockey team, a gaggle of basketball players and a number of fighters. Of these latter he took particular pride in Lee Dockerty who had originally caught his eye in an amateur tournament and who he had supported all the way to his winning of the world's middleweight championship.

While he administered his own stud farm and raced under his own name and colors, this being a respectable rich man's sport, the other interests he turned over to the management of what amounted to the Mafia's Eastern Sports Department and in the main he left the operations to them, provided on the whole they delivered the goods.

When from time to time strange things happened in the name of Uncle Nono, Gio rarely gave signs of displeasure. If a football team mysteriously lost a game it should have won; if there were rumors that a certain out-of-town engagement between a couple of bruisers had been a waltz; if a basketball scandal momentarily rocked the game or seekers for information along the Great White Way would let slip out of the sides of their mouths, 'Which way is Uncle Nono betting?' no one would connect any of these with Gio di Angeletti. But this affair of Lee Dockerty and a ridiculous animal by the name of Matilda, which might result in the impalpable figure of Uncle Nono being laughed at, was something else again.

For there was a curious ambivalence of vanity here which often worked to the disadvantage, sometimes permanent, of those who let him down and this ambivalence was the result of the hidden duality of Giovanni di Angeletti and Uncle Nono. For even though, with the exception perhaps of certain duplicate files in the offices of New York's District Attorney and the local headquarters of the F.B.I., the work of dedicated researchers who had still not uncovered enough to go on, there was no hint or knowledge, or more than a vague guess at the identities of the two names.

But the fact was, nevertheless, that if you hurt Uncle Nono, you bruised di Angeletti. If it became known along the Great White Way that Uncle Nono had in some manner come a cropper, the pain also registered in the afflatus of Gio. To him and, of course naturally, to top members of his secret empire, the two were inseparable. Hence steps had to be taken.

Di Angeletti, sitting behind his huge, ebony, glass-topped desk, looked like what he was where he was, a perfectly groomed businessman in expensively conservative clothing. He had the Sicilian family nose, a powerful beak and those heavy, slightly moist, intelligent Italian eyes shining above it. He was six feet in height, weighed a hundred and ninety-five pounds—only five more than he had scaled as a halfback at Harvard—and wore his still dark hair in a wedge-shaped forelock that fell over his right brow. His voice was gentle, his movements clean-cut and decisive.

The door opened admitting Johnny Renato, smart and dapper, carrying his tan topcoat over one arm and his jaunty hat in his hand, and at that moment Gio ceased to be di Angeletti and became an irritated Uncle Nono who motioned his second-in-command to sit, opened the desk drawer and took out copies of the *Mercury* containing the offending articles by O'Farrell and Parkhurst.

He said, 'Well!'

Renato replied, 'Okay, chief. It's all arranged. I don't think you'll be bothered any more. But when this is cleared up we'll probably want to make some changes. Pinky is a boob and Joe Marcanti isn't much brighter. We'd be better letting a man like Paddy Aloysius manage Lee Dockerty,' and then he added, 'that's if he's still around.'

Uncle Nono looked again at the paper that contained Duke Parkhurst's lead article and also the final edition of that morning's *Mercury* with O'Farrell's story and the mocked-up photo montage of Lee Dockerty squared off against Matilda. As always, di Angeletti had difficulty in keeping the corners of his mouth from twitching. It had been a damned funny story and equally damned bad luck that Parkhurst had been there to cover it.

Gio di Angeletti wanted to laugh, but Uncle Nono could not afford to do so. He dared neither look nor feel amused in the presence of those most humorless of men, his brother Mafiosi who took themselves and their positions with a seriousness which led to an easily affronted dignity.

The ambivalence set in again. Duke Parkhurst was the real enemy who was threatening to make a laughing stock of him. His conceit reasserted itself. If other readers did not know that Lee Dockerty was the boy of that shadowy, unknown, unidentified figure referred to as Uncle Nono, Parkhurst's following did. At the Friday night fights at the Metropolitan Arena there would be jokes and gags passed between the insiders at the expense of Uncle Nono, and the snickers would cut through and penetrate to di Angeletti's secret ego.

For a moment through his mind there coursed that line of Henry II to his courtiers, with reference to Thomas A' Becket, 'Will no one rid me of this turbulent priest?' He knew he had only to paraphrase it to Renato and before too long Duke Parkhurst would meet with an accident. His Sicilian heritage versus his American birth and education fought out a brief duel and he resisted the temptation. It was not worth the risk of destroying a billion-dollar empire for one moment of Mafian vengeance. His victory over himself lay in the fact that to murder, or to have murdered, a sportswriter was unthinkable. It was un-American and Gio di Angeletti had always tried to run his American section in accordance with American *mores* and justice.

Renato's eyes had followed Uncle Nono's preoccupation and his intelligence practically read his thoughts. He said, 'Maybe just a broken leg?'

Di Angeletti was not even tempted again. He shook his head and said, 'No, Johnny, no rough stuff. But maybe there are other ways.' For in front of his second-in-command his bruised vanity had taken over again. Parkhurst, obviously working in the dark and based on Broadway gossip linking Uncle Nono to occasional old happenings in the world of sport, had been indulging in isolated digs. But this affair of Lee Dockerty and his middleweight championship was more like a concerted, continuing attack, and wanted stopping

110

before it might spread. He said, 'I hope these fellows of yours will work with discretion, Johnny. Are they really smart?'

Renato said, 'Sure, chief. Joe Marcanti says they've done some good jobs for us.'

Uncle Nono, now turning back into di Angeletti once more, nodded. But he was not feeling too certain even though he had okayed the steps that Renato had recommended and then put into motion. One of his problems as an educated and sensitive man who was judge, jury and executioner all rolled into one, was that to carry out his duties he was forced to rely upon help that was certainly not in a class to be compared with him intellectually, and even definitely dim. It was one of the rather insoluble facets of his job. The educated, subtle, intellectual killer-for-pay simply did not exist. You had to use what you could get. Aloud he said, 'I wouldn't like Parkhurst to latch onto something he could really get his teeth into. But then maybe we might get there first. What's his weakness?'

'Honesty,' Renato replied. 'In this racket it's likely to get you into more trouble than anything else.' And then, as an afterthought of not too great importance, he added, 'They say he likes girls, but you can't sue him for that. What I mean is not a chaser and not ever in any trouble. Married some society dame and they were divorced last year. Nothing you could put your finger on—the decree didn't even say cruelty, but incompatibility. Since then he hasn't had any regular; you know, sort of playing the field, but more of a loner. Eats out a lot by himself or maybe with some of the other sportswriters.'

Gio took a pen from an onyx holder and began to play with it, outlining Matilda's head and stand-up ears in the newspaper picture in red ink. Then he asked, 'Is the Mann Act still operative?'

Renato looked startled, 'They haven't pulled anyone on that for years, but I think it's still on the books. We're not going back into that business, are we?'

Di Angeletti shook his head in the negative and then said, 'Check it, anyway.' Then he added, 'By the way, where's Irresistible?'

'You mean Birdie?'

111

Gio nodded, 'Sweet kid. I haven't seen her for months.'

'She's around. She's still got that dancing part in "Hello Soldier", but she's been featured since—since you took an interest in her.'

Gio made a note on a scratch pad on his desk and again murmured, 'Sweet kid. They don't call her Irresistible for nothing. She might be ideal, she owes me a favor.' He looked up from his notation and said, 'Thank you, Johnny, that's all.' And then casually, 'Incidentally, have a cheque for a hundred thousand dollars sent to the *Daily Mercury* Free Food Fund for Hungry Children, from my personal account.'

Renato looked at him in amazement.

Gio di Angeletti smiled benignly, 'That will be their lead story *tomorrow*,' he said.

Renato had the exit line, 'Chief, you're the cleverest!'

# XII

PATRICK ALOYSIUS HAD scheduled Matilda's build-up itinerary to proceed south-westwards out to the Coast and then jump back to try to take in some of the larger eastern cities.

They were in Clayton, Oklahoma, to engage a local product known as Cowboy Jones when the deputation despatched by Joe Marcanti arrived and filed into the premises, that Ahearn, Bimmie and The Bermondsey Kid were occupying in the Cattleman's Rest Hotel.

As a thriving cow town of some 10,000 population and just across the river from the Texas Panhandle and a smaller community from which it drew customers, Clayton was able to afford an auditorium known as the Junior Cow Palace Arena that could support boxing, wrestling or be converted for its championship, high school basketball team. The citizens of Clayton and vicinity were highly sports-minded and fascinated to find out how the Cowboy, who came from the Panhandle with a fair to middling reputation, would make out against 'that there kangroo'. They were a betting crowd and since Clayton also entertained a biannual rodeo, they were no strangers to contest between man and beast and wagered large sums of money on calf roping, bronco busting, steer wrestling and Brahma bull riding. A kangaroo was something new to them in the animal line particularly one that stood up on its hindlegs and boxed, had knocked out the middleweight champion of the world and had already added several lesser known professionals to his list. On the other hand in that part of the country man usually emerged triumphant over beast in the various events of the rodeo. All this threw the odds makers into a considerable state of confusion and one could bet it any way one wanted. This in no wise interfered with the enthusiasm with which the forthcoming encounter was awaited.

The three visitors entered the room without announcing themselves or bothering to knock upon the door. Bimmie, Patrick Aloysius and Billy Baker were playing gin rummy to while away the afternoon.

Matilda was bedded down comfortably in the stables behind the hotel. If kangaroos were capable of happy anticipation, he was enjoying such for he seemed to know either by instinct or the turn taken by routine that he was due for a contest in the evening and this always pleased and excited him.

In their Broadway clothes, violently coloured shirts with equally flashy neckwear and the fashionable peanut-sized, almost brimless fedoras the First, Second and Third Murderers could not have been more conspicuous as strangers in town if they had arrived clad in tutus, but this did not seem to worry them. The names of the three hoods despatched by Marcanti were Pietro Delzi, Salvadore Ciampanza and Alfredo Gentile.

Delzi was known as Nickel-Plate from his custom of using a nickel-plated .32 to carry out his assignments. He had no need to be a good shot since it was his habit to place the little gun against the spine of the offender before pulling the trigger. Ciampanza naturally was known as Chimpanzee, or more familiarly Big Chimp. He was a large, gross man who considered himself something of a comedian as well as a specialist in rapid and accurate fire. His executions always bore his signature—five bullet holes neatly grouped within a radius of a silver dollar piece.

Alfredo Gentile was a small, soft-spoken man with a face shaped like a skull who wore his striped, collar-attached shirts so tight around the neck that they seemed to be choking him. His nickname was Smiley, derived from his death's-head grin that accompanied the introduction of a thin-bladed stiletto between the third and fourth ribs at such times when silence was considered prudent.

All three seemed to have entered the room simultaneously and stood in the doorway. Big Chimp said, 'Hullo boys! We just thought we'd drop in. Who's ahead?'

114

Patrick Aloysius had been in the act of picking up a card, but his hand froze in mid air. His eyes shifted uneasily and he said really unnecessarily, 'Hello, fellers. Come in,' but did not pick up the card.

Bimmie had never seen any of them before but the atmosphere of menace penetrated to him and he looked anxiously at Patrick Aloysius.

Baker asked, 'Who are they, friends of yours?'

Smiley Gentile, appearing to gag in his collar, said in a soft voice, 'Go on. Finish out the hand, boys. Don't mind us.' They found themselves chairs and sat down negligently with their legs crossed and without removing their hats.

Baker said, 'They ain't got very good manners, 'ave they?'

Out of the side of his mouth Patrick Aloysius said, 'Shut up!' His usual flush had drained away leaving his face as white as his hair, since he did not yet know the purpose or the mission of Uncle Nono's execution squad.

Nickel-Plate produced his .32 from one pocket, a silk handkerchief from another and proceeded to polish his weapon with the concentration of one who likes his artifacts immaculate. Big Chimp was toying with a flat automatic. Smiley sat grinning mirthlessly with nothing in his hands, leaving his personal choice of hardware momentarily to the imagination.

Patrick Aloysius managed to continue the operation of picking up the card, found that it filled, laid down his hand and said, 'Gin!'

The Chimpanzee said, 'You always was a lucky bastard. I oughta cut myself a piece of you, except all your fighters are bums.' His horn-rimmed glasses gave him the aspect of a lawyer rather than the Second Murderer. Nickel-plate was younger, one of those blond Italians with contrasting dark, liquid eyes.

Patrick Aloysius pushed his chewing tobacco far up into the top cavity of his left cheek to eliminate any danger of swallowing it. His voice was steady enough now as he asked, 'What can we do for you boys?'

Smiley nodded pleasantly, leaned forward, his voice reduced to almost a whisper, 'Just a little chat. No trouble or anything like that. We know you'll like to go along with us. It's just that we think your boy . . .' He gagged here both from his collar and the word 'boy' which suddenly took him short.

Big Chimp laughed, 'Boy is good!'

The First Murderer recovered himself and continued, '... well, whatever it is, he don't win tonight. In fact we don't think he win at all any more, see?'

Bimmie said, 'Whaddyer mean, we don't win tonight? We ain't selling out to anybody.'

The First Murderer shook his head gently and sibilated, 'No, no, no! You don't understand. You ain't being asked to sell out. There's no dough involved in this, it's just if your ...' He hesitated again, '... your fighter would pick out a good punch maybe in the fourth round—he don't have to get himself hurt—just so it looks good—and stays down for ten seconds after which maybe you could announce his retirement from the ring.'

Patrick Aloysius had recovered some of his color and in fact red was beginning to climb up out of his open shirt neck. He said, 'What if we don't?'

Smiley twitched his death's head in the direction of the Second and Third Murderers. He said, 'It might just make Nickel-Plate and Big Chimp here irritable. We'll be at the ringside. You wouldn't want a lot of noise there, would you? Speaking for myself personally, I don't like noise.'

He had hardly moved his fingers during this soothing speech but as if by magic they were now caressing the long blade of a stiletto.

Up to that time Bimmie's work had not taken him into the ken of the Broadway underworld. He had been much too small fry to attract the attention of that type of businessman or the protection racket. He felt a surge of honest rage and had the courage to voice it. He burst forth, 'What, you mean, you'd risk a murder rap if our kangaroo don't lay down in the fourth?'

Smiley's voice became even softer if possible as he sighed, 'No, no, no! We don't want no murder rap if we can help it. You ain't looked up your law books, Sonny. There ain't no murder rap in Oklahoma for shooting a kangaroo.'

Billy Baker cried, 'Blimey! And you call this a civilized country?'

Big Chimp chuckled and said, 'Oh, I dunno, Limey. Our

connections over there tell us you boys ain't so bad yourselves. Now them Kray brothers of yours could have gone far if they hadn't lost their tempers and got careless. Take us—we never lose our tempers and we never get careless.'

Patrick Aloysius had a try, 'Can't we work something out?' he asked. 'There's been a lot of money bet on this match.'

'Yeah, we know,' Gentile said. 'But see, it's got to be like we want it.'

Nickel-Plate had finished shining up the snub barrel of his .32, gave it a triple flip in the air, caught it and returned it to his pocket.

Bimmie was shattered to a state almost of tears as he saw and understood the menace of the three torpedoes as the end of the road to riches for them all. He could not even think of a line of schmoos but shouted, 'Why don't you bastards go down to the stable where he is, then, and shoot him now?'

Smiley proved the point that they did not lose their tempers when he disregarded the epithet and said, 'Why, that would be downright wasteful. We been around town a bit; one will get us ten, if we name the round in which the Cowboy knocks out your thing. Nobody minds us making a bit on the side as long as it's honest and doesn't interfere with orders. Make it the fifth round, say, when he lays down and that will give the crowd a run for its money. Be seeing you.' They arose simultaneously as though jerked by one marionette string.

Patrick Aloysius said, 'Hey, wait a minute! How do you expect us to tell a kangaroo he's to go into the tank in the fifth from a slap on the elbow?'

Big Chimp enquired, 'Don't they speak English down there in Australia?'

'Sure they do, but our kangaroo don't. He don't speak anything. He don't understand anything either, except when he sees a fella standing up against him wearing mittens; he wants to...' 'Murder' did not seem quite the appropriate word here, so Patrick Aloysius settled for, '...chop him down.'

Smiley's grin more than ever showed the skull behind it.

117

He said, 'Maybe you'd better speak to him in kangaroo, then. Anyway, as I see it, boys, it's your problem. If the kangaroo don't go like we said, maybe you do.' They exited without bothering to close the door behind them.

The three men left sat around the card table silently for a moment. Patrick Aloysius mopped his face, spat out his plug of tobacco and said to Bimmie, 'Your friend Parkhurst has got under Uncle Nono's skin.'

Billy Baker said, 'Then why don't they go and shoot Parkhurst?'

Patrick Aloysius replied, 'There isn't any open season on sportswriters. Fellers like us, if the occasion arose, might be considered more expendable.'

Bimmie said, 'Can't we tell the police? We got our rights.'

Patrick Aloysius's expression went sardonic again, 'You got your rights but this wouldn't be the moment to trot them out for exercise. You'd get laughed out of town. Tell 'em what? That three hoods have come down here to shoot up a kangaroo if he don't do a splash?' He suddenly turned his deep-sunk eyes on Billy Baker and said, 'Can't you get through to your animal and get him to take that dive just this once? One kayo-by on his record won't hurt him none. Most fighters catch at least one on their way up. He'd understand if you explained it to him, until we can see what to do.'

It was the little Cockney's turn to be ruffled and angry. 'He wouldn't do it, even if I could explain to him. How would I go about telling him? He just knows his nyme and that's all, besides boxing.'

Bimmie said, 'Those bums can't do that to us! I'll wire Mr Pockhurst. We'll call off the fight. Who the hell does this Uncle Nono think he is?'

Patrick Aloysius raised a cynical eyebrow, 'Who's Uncle Nono? It isn't considered healthy to ask. Ain't you never heard of the Mafia? Grow up, Bimmie, you ain't booking Chink magicians and half-ass ventriloquists any more. This is a tough racket you're in now. It's too late to get hold of Parkhurst and if you call off the fight they'll just knock off Matilda anyway. The only reason they ain't done it yet is because they're greedy. Those odds are what saved him. We got to think of something else.'

118

Billy Baker arose and headed purposefully for the door. Bimmie said, 'Hey! Where do you think you're going?'

The little Cockney with the punched-in face was ablaze with wrath, 'Where am I goin'? Out to buy a gun! Nobody's goin' to shoot up my 'roo and get away with it!'

On the drive from the station to the hotel The Bermondsey Kid remembered having seen a fishing tackle, gun and sporting goods shop a few blocks from the Cattleman's Rest and this was his destination. But first he went to the stables behind, relic of the days when Clayton was a frontier town and where Matilda had been comfortably installed. When he saw The Kid come in, he stood up on his hindlegs and they embraced. The trainer pulled the big, forward-looking ears, scratched him between them, endured half a dozen soppy licks and said, 'You just go in there and zap that Cowboy bloke and anybody tries anything funny with you will get his bloody guts blown out. Here. Here's a down-payment for you and there are more to come if you zap him in the first round,' and he unwrapped and handed Matilda a Hershey Bar and departed followed by the kangaroo's grateful, 'Uck-uck-uck-uck-ucks!' muttering to himself as he went, 'Bloody Yanks! We'll show 'em the empire ain't dead yet!'

In the gun store The Bermondsey Kid marched to the counter where a variety of weapons were on display and said, 'I want to buy a gun.'

The salesman, a middle-aged, leathery Oklahoman said, 'Yes *sir!* Glad to oblige. Rifle, shot or hand?'

'A pistol,' said Baker, 'something reliable.'

The salesman said briskly, 'Reliable! That's exactly the word you wanna think of when you're buying a gun. Any special calibre?' Before his customer could reply and having sized him up, he said, 'You wouldn't be interested in these here ladies' models, "Husband Specials" we call them, ha, ha!' and steered him away from the display of .22 and .32 calibre articles. 'A thirty-eight's a nice size. A forty-five blows a bigger hole but they like to throw up high if you ain't used to them. Automatic or regulation six-gun? Lot o' folks say them automatics is likely to jam on you, but the way they're making 'em today, they're pretty good. Now take this here Colt, for instance. I never had nobody ever come back and say they'd had any trouble,' and he produced a wicked, black, flat .38 calibre, short-barreled automatic. 'Fits the hand nicely,' he said, 'but then, you know how fellas around here are, they're hung up on the good old six-gun because that's what their pappies used to pack in the old days, and they think it's sissy to use them automatics or semi-automatics—might as well use a machine-gun and when you start spraying lead around with one of them things you could just pick off some innocent bystanders, which nobody wants on his conscience. Now if I could recommend something guaranteed to give satisfaction, can't jam, easy on the trigger and shoots straight as a die...' He produced a beautifully blued, .38 six-shooter with a medium-length barrel and a rough horn butt with an especially carved platform on the side for an easier thumb grip. 'If you just take the tip of the sight and squeeze, you can't miss. Sixty dollars.'

'I'll 'ave it,' The Bermondsey Kid said, and reached for his wallet.

'Will you want a holster, too?'

The Bermondsey Kid reflected. If he had to use it quickly, a holster would only get in the way. 'No thanks,' he said, 'I'll 'ave it like it is.'

'Fine!' said the salesman, 'You a resident?'

'No. Just passing through, as you might say.'

'Well then maybe you've got a note from our sheriff,' the salesman suggested.

The Bermondsey Kid halted midway extricating the sixty dollars from his bill fold and said, 'What? Note from who? I ain't got any note.' And then he added from what he had gathered by osmosis, as it were, since his arrival there, 'I thought in Oklahoma you didn't need any licence to buy or carry a gun.'

'That's right, that's right!' the salesman agreed. 'Nor you don't if you live here. This is a civilized place where a man can be trusted. Only you see, when someone we don't know in town comes in and wants to buy one, the sheriff maybe just likes to know what he's planning to use it for.'

The ways of the Yanks were evermore bewildering to Billy

Baker and he said, 'I don't get it. Either there's a bloomin' law or there ain't and, what's that got to do with the sheriff?'

'Well now, you see,' the salesman soothed, 'it ain't all that complicated. Like I say, the law's there to protect everyone and if you know the other fella's got a right to pack a gun, maybe you won't get so shirty starting arguments. But now, supposing some lady gets a mad on at her husband and thinks she could maybe solve her problems with a little homicide . . . If the sheriff knowed about it, maybe he could talk her out of it and save a lot of trouble later. Or, say some fella suddenly got the idea somebody had done him an injury, well, a little chat with the sheriff might calm him down. You'll like the sheriff, he's a fine fella; name of Sam Baker. Tell him I sent you. You'll be back in no time with a note and I'll have it all wrapped up for you.'

The Bermondsey Kid wanted that gun. He nodded, went out, crossed the street and down a block as directed and entered a store front that had 'WATAHATCHIE COUNTY OFFICE OF THE SHERIFF' lettered on the window.

Two deputies were lounging in the front portion, which was decorated with handbills showing photographs of various miscreants and describing what they were wanted for, and their eyebrows went up at The Bermondsey Kid's accent when he asked, 'Where can I find your 'Igh Sheriff, Mr Baker?' They nodded their heads in the direction of a rear office where that individual was seated at an old-fashioned, roll-top desk doing his paperwork.

He looked exactly the way The Bermondsey Kid thought a sheriff ought to look, except for the wearing of a pair of steel-rimmed spectacles, a huge, lanky man cut out of rawhide, with a long horse face whose chief feature seemed to be a pair of bushy and permanently amused eyebrows over a pair of penetrating, far-seeing eyes. On a peg behind the desk hung a sombrero and The Kid, who had seen his share of Westerns, was glad to note that it was white and not black. Also suspended there, encased in a leather holster with cartridge belt, was the largest hip cannon on which he had ever set eyes, in or out of the films.

The sheriff looked up from his work and said, 'Good afternoon, suh,' in the classic Western drawl he might have

121

been expected to use, 'in what manner or way might Ah be able to serve yo?'

'I want to buy a gun,' The Bermondsey Kid said.

'Sho', sho', said the sheriff, 'That'll be all right. Ah reckon yo' all are the fella in charge of thet there trained kangaroo that's fightin' down at the Arena tonight. Ah trust none of our citizens hev bin lacking in hospitality or failing ter make yer feel right ter home in our midst?'

The Bermondsey Kid remembered what Patrick Aloysius had said about the uselessness of informing the police about the dilemma that had been set them, so he just reiterated, 'I'd just like to buy a gun.'

'Sho', sho'!' repeated the sheriff soothingly, but the steel-blue eyes beneath the quizzical eyebrows were taking in the little banty ex-fighter, 'Wa'all now, we'll just write a little note to Bill Powell, down at Powell's store and he'll sell you one.'

The sheriff took a scratch pad, poised a biro and said, 'Name?'

'Billy Baker.'

'Home town address?'

The Bermondsey Kid hesitated. It had been so long since he had been there; he had roamed so far and wide that he could hardly remember the address any more. But the sheriff took his hesitation for some kind of reluctance and looking over his spectacles said, 'Jes' a formality, suh, where yo' all would like to be delivered in case of an accident. Guns have a way of being mighty sudden when there's mo' than one party involved.'

The Kid finally remembered, 'Fifty-eight Windmill Rise, Rotherhithe Street, Bermondsey, London, England,' he said, 'I've got an Uncle George living there, if he's still alive; I wouldn't know.'

The sheriff had swivelled round, leaning forward in his chair for a better look at his visitor. Almost unbelieving he was saying, 'Git out! Did I hear y'all say Rotherhithe Street in Bermondsey? Well, whut d'yer know! I cain't hardly believe it!'

'Believe what?' The Bermondsey Kid asked.

The sheriff was looking at the name and address he had

122

written down on the sheet of paper and said, 'Why, thet practically makes us kin folk.' He suddenly stuck out a huge hand at The Kid and said, 'Sam Baker's the name. Mah great-gran' pappy was bohn there and so was the gal he married. Thet part of the town was full of Bakers. He came to the Indian Territory—that's what Oklahoma was called in them days—way back in 1880. He took part in the Great Homestead race of 1889, got himself a passel of land and us Bakers have bin here ever since. But we was as British as yo.'

Another light dawned upon the sheriff, raising the eyebrows still further. 'Why, you'all must be the Billy Baker who I read about when I was a young man, who was lightweight champeen of England, The Bermondsey Kid. Well-now, son, if this ain't a mighty proud moment!' He enfolded The Kid's battle-scarred hand in his big paw and was shaking it. He became suddenly aware of the slip of paper he was still holding in his other hand and now he looked at it in a puzzled fashion, broke off and said, 'See here, Cousin Billy—'cos I'm bettin' yo' ain't no less—what yo'all want thet there gun for?'

And this time The Bermondsey Kid told him in lengthy and descriptive detail.

The sheriff sat silently throughout the recital. When The Kid had finished he shook his head gently from side to side and said, 'They sho' is a lot of funny people around in this world, ain't they?' Then slowly he tore up the paper and dropped the pieces in the waste basket. 'Cousin Billy,' he said, 'you ain't goin' to need no gun as long as Ahm sheriff of Clayton. Save yo' money. I got all the gun anybody ever needs.' He rose and lifted the cannon that had been hanging on the wall behind him and placed it on the desk with a crash. It was an old-type frontiersman Colt, five-chambered revolver, but bigger than anything The Kid had ever seen. The sheriff said, 'You know what thet is? Thet is a fifty calibre and kin knock down the side of a house. It belonged to my great-gran'pappy. He called it Old Smokey. Mr Colt made only about twenty of 'em before he hit upon the forty-five for standard as being easier to heft and draw. But mah great-gran'pappy was a big man, bigger than I am.' He pulled open a drawer in his desk and there was a rattling of ancient

123

brass pistol cartridges which looked to The Bermondsey Kid as big as wartime, six-millimeter cannon shells.

The sheriff thrust five of these into the chambers, gave the cylinder a whirl and thrust the weapon back into the holster. Smiling benignly he said, 'Jes' you go on about your business, Cousin Billy, and leave everything ter me.'

The affair at the Junior Cow Palace Arena was brief, rousing and to all present, with a few exceptions, highly satisfactory. Whether Cowboy Jones, a heavily muscled fighter who wore his hair long and looked more than 160 pounds, had been apprised that he was to triumph in the fifth round and hence failed to exercise due care, nobody ever found out. After two minutes and six seconds of the opening inning, Matilda's whip punch caught him on the side of the jaw and deposited him unconscious in his own corner, to be counted out while Matilda, a happy kangaroo, retired to the prescribed neutral area and was congratulating himself with his paws clasped aloft in the gesture that Billy Baker had taught him.

As the referee, whose arms had been waving up and down for the count, flailed them across the body in the ten-and-out gesture, three strangers at the ringside arose and the cluster of arc lights under the canopy overhead picked up the silver glitter of something brightly polished.

Simultaneously Sheriff Baker was on his feet, Old Smokey in his fist pointed to the roof, and pulled the trigger.

Inside the enclosure the ancient weapon went off with the shattering roar of a six-inch gun and a cloud of black smoke not unlike the atomic one in shape and effect. It had the immediate result of stunning every person there into the momentary immobility of total shock, including the three strangers, two of whom had guns and the third a knife in their hands.

Bits and pieces of the injured roof which had a hole the size of a grapefruit blown through it, came drifting down as the sheriff addressed the newcomers mildly.

'We'll hev' them thar' toys, if you don't mind, gentlemen,' the sheriff said. Whether it was still the effect of the explosion of Old Smokey or the fact that the sheriff was using the great cannon to gesture with as he directed his two deputies, the three gave up their weapons meekly.

The sheriff ordered his men, 'Okay, boys, take 'em.' The deputies produced handcuffs and the sheriff tossed them a third pair of his own.

Only then did Smiley Gentile unfreeze long enough to gag out the words, 'What's the charge, sheriff? We ain't done anything to nobody. We were only after the kangaroo; he might have killed that boy.'

'Yeah, Ah know,' said the sheriff. 'So Ah heard. The charge would be huntin' without a licence. That's a serious offense in our County. We're mighty careful about protectin' our game. Jedge Harman will likely call it thirty days and five hundred dollars' fine.' Then to the deputies he added, 'Okay, fellas, take 'em to the jail. I'll make the charge official later.'

After this the pent-up emotion and the excitement of the crowd burst in congratulations, accolades and back-slapping for the sheriff, Bimmie, Patrick Aloysius and Billy Baker, while Matilda took it all personally, still clasping his paws overhead.

The sheriff said, 'Cousin Billy, now we'll all go down to my house for a little celebration, and don't forget to bring Matilda. Ah've promised the kids. Ah've ordered a couple o' quarts of chocolate ice cream for him.'

This was essentially the way Duke Parkhurst wrote up the inside story the following day, having heard it from Bimmie over the telephone.

Uncle Nono was never mentioned, of course, but his evident irritation was manifested a few months later with Mr Gentile who instead of carrying out his original instructions to eliminate either Matilda or his handlers or the lot, in case they should not be amenable, had permitted his compulsion for a sure-thing gamble to wreck the project. He was eventually discovered encased in concrete, blocking one of the drains of the reservoir at Pocatello, Idaho. And Uncle Nono, having had his little chat with Birdie, now ordered the immediate implementation of Plan B.

# XIII

SUNDAY EVENING WAS the night that Duke Parkhurst took off from his usual Broadway haunts, passing up the steaks of Toots Schor or the ravioli of the Taverna for the *sauerbraten* or goulash with *spätzli* and *Spatenbrau* at Luchow's famous, almost century-old German restaurant on 14th Street, where the theatrical crowd was wont to foregather. He liked German cooking and the place was a rest as well from the eternal yaketting about football, baseball, boxing and all the other muscle games. He always had the same table by himself where the celebrities were placed, commanding a view of the entrance. There, he could sit quietly midst the heavy German cookery fumes and recharge the batteries, while the six-piece string orchestra scraped away at potpourris of 'Fledermaus' or 'The Chocolate Soldier'.

There is a foyer at Luchow's leading from the street to the actual entrance to the restaurant and on Sunday night the velvet rope goes up to act as a barrier and hold back the mob not only of would-be diners who have neglected to make a reservation, but also the gapers who come to stare at the famous.

Parkhurst, looking over the crush of outsiders, was aware of a pretty girl—a face, a form, hair, attractive clothing. He also, at the same time, noted a man standing close to her he might have seen somewhere before.

If Parkhurst had been thinking of applying words to the tiny wisp of emotion caused by his glimpse of the girl jammed in the crowd in the entry, he would have written 'appealing', for this was its message, its aura, not broadcast

126

directly but simply there; something youthful, large-eyed, tremulous and pleasing.

At this point a table-hopping actor buttonholed him and when finally the intruder had disgorged his accumulated platitudes and taken himself off, the girl was gone. Parkhurst had no idea whether she had passed the barrier, been shown to a table somewhere in the rear, met a date who had been waiting for her or, discouraged, had simply left. But since she had vanished he sensibly put her out of his mind, ordered a fresh tankard and gave himself up to the nostalgic relaxation of the string orchestra's *schmaltzy* rendition of excerpts from 'The Merry Widow'.

'Vilya, Oh Vilya, the witch of the wood.'

And then suddenly she was standing beside him at his table, looking down at him with startled eyes and a worried expression, the attitude of her body half poised for flight. Parkhurst was at once aware of the return in full flood of his original emotion. There was something extraordinarily touching about her. She was clutching a small, leatherbound autograph book.

She laid her hand upon his arm, an instinctive and unconscious gesture and said, 'Please don't get up. I feel awful disturbing you. It's terrible of me. Could I ask you, please, to sign my brother's autograph book?'

Parkhurst said, 'Of course. Sit down, won't you?'

She did so, but only on the edge of a chair and he took the autograph book from her and leafed through the pages. Now that she was there, he did not want her to go away. He had not yet had time to study her features and to see what it was at first glance had caused that astonishing small but continuous tug at his heart.

On the flyleaf in a still unformed hand was scrawled, 'Bobby McBride, aged fourteen. DeWitt Clinton High School. N.Y.'

The girl saw his glance and said, 'I'm Birdie McBride. I'm...'

The 'Birdie' connected with 'McBride' now jogged Parkhurst's memory and he smiled and said, 'Of course! The girl who stops the show in 'Hello Soldier.' I liked your dancing. It's most effective. It made me feel sad and I think

127

perhaps others in the audience too. I saw a lot of handkerchiefs come out at the finish.'

'Yes,' said Birdie, 'I think sad thoughts when I'm dancing it. You see, my boy-friend, this soldier with whom I've been spending the weekend, is going away and I know he isn't coming back because I've found out he likes the other girl better, and that's why my dance is sort of sad.'

Parkhurst looked at her sharply, for she had been quoting the scenario of a show he had seen, but he observed to his astonishment that she was quite serious and even introspective. He asked, 'Where do you come from, Birdie?'

She replied, 'Milwaukee.'

He asked, 'Is your family there?'

She said, 'My Mum and sisters still are. Daddy's dead.'

'Have you been in New York very long?'

'Only about two years. I went to dancing school in Milwaukee and they said I ought to try out. Mum let me come to New York.'

Parkhurst nodded. The mystery was becoming intriguing. He re-examined the autograph book. There were the usual signatures of ball players, professional football heroes, fighters, actors, an author or two; each page was a different pastel color.

Parkhurst suddenly looked up from his inspection, 'May I buy you a drink? Are you alone, or are you with someone?'

Birdie said, 'I oughtn't to. I'm taking up your time.'

Parkhurst said, 'It's bearable. What will you have?'

He saw her glance fall swiftly on his tankard before she replied, 'Oh, I'd love a beer.'

Parkhurst nodded, called the waiter and ordered a glass of the light Pilsener. He chose a pink page as one that suited the occasion and inscribed it, 'To my friend Bobby McBride, with all good wishes. Duke Parkhurst,' and handed the book back. The waiter brought the beer. For a moment, as they lifted the golden brew to one another their eyes met over the rims and she smiled shyly and gently. After the first sip she opened the autograph book and Parkhurst saw that for an instant her glance lingered upon the flyleaf page with her brother's name and then she turned quickly to his own inscription. 'Oh, thank you, Mr Parkhurst.' She ran the tip of her finger over the signature as though to gain closer contact. Everything she did was utterly charming.

It was also, Parkhurst thought, as blatant a pick-up as he had ever experienced. What made it interesting was that the girl was not some unknown little tramp looking for a night's stake, but a solo dancer in a Broadway show whose bit, though brief, had attracted critical attention and who might be said to be standing on the threshold of a career.

And now that she was sitting there quietly opposite him, clad in a gray frock with something white and frilly at the throat, completely oblivious of the fact that he had penetrated the fraud of her approach, he had an opportunity to try to evaluate her and see what it was that had so struck him the first time he had picked her out of the crowd. To have attempted to describe her, as Parkhurst afterwards said, was simply to put together a conglomeration of clichés; large trusting doe's eyes in a child's face, a body as slender and pliable as a young sapling, a hank of light brown hair softer then silk, worn in a twist high on her head so as to reveal the curling tendrils on the nape of her neck, a tremulous mouth over fine teeth, dainty ears, delicate blue veins at her temples...

None of these features were those of what might be termed a great beauty but they added up to something else and whatever it was, it clutched at his heart. She was...he had the word now, irresistible.

Common she probably was, in the sense that her speech showed she had not progressed too far in her education; common perhaps but not vulgar. She had probably come up from the chorus before she had been given that bit to do in the show, but she lacked that brassy showgirl quality. Parkhurst had a feeling of restfulness with her. She sat sipping her beer and did not speak until she was spoken to. This was the second mystery connected with her and Parkhurst felt compelled to solve it before he tackled the first. It was one thing to fall for a pretty face; another to be strongly moved by someone he had not known until a few moments before. He asked her, 'Have you ever been here before?'

'Oh no, never.'

He nodded in the direction of a table four removed from them and asked, 'Do you know who that is over there?'

Birdie shook her head in negation.

'That's Marlon Brando.'

'He's nice-looking.'

'And that's Peter O'Toole and Leonard Lyons sitting with him.'

Birdie took them in with her soft gaze but nothing was happening. She asked, 'Are they all famous? As famous as you?'

Parkhurst asked, 'What makes you think I'm famous?'

Birdie did not answer but only smiled gently at him.

Parkhurst began to talk now. He tried her on many topics to which Birdie only listened silently and without comment. He talked about himself, his work, the state of the theatre, the new permissiveness, the changing world, the insensate rebellion of the confused young. From time to time a flicker—but a flicker only—would pass across her enchanting façade to confirm that for the most part Birdie was not with it. A conviction that had been a suspicion earlier in the evening was now confirmed; Birdie was not very bright. Parkhurst now tackled the second mystery.

He reached over, picked up the autograph book again, tapped it slowly against his open palm and asked, 'This isn't your brother's book, is it?'

She put down her glass and looked at him in amazement. 'No,' she admitted. 'How did you know?'

The complete innocence of her demand made Parkhurst ashamed at how easily she had fallen into the trap he had set for her. The phoney book and the honest dancer—what was the connection? And then clear as a television beam, up from his subconscious drifted the individual who for a moment he had seen standing close to Birdie, thin-faced, with drooping nose and drooping lip rather like a camel, and he remembered then where he had seen him before—once at a Lee Dockerty training camp. The finger man!

He asked, 'That fellow who was with you when you were standing over there in the entrance, he was there to point me out to you, wasn't he?'

Birdie nodded again. She replied, 'Oh, did you see him?

He wanted to be certain I didn't make a mistake.'

'And after he pointed me out to you, he left?'

'Yes.' And then with that completely disarming simplicity she asked, 'Are you angry with me that I made friends with you?'

Parkhurst replied, 'No, I don't think so. I don't think I'm angry at all.' And then he added, 'The fellow who pointed me out, he wasn't the one who sent you, was he?'

Birdie looked horrified, 'Oh no!' she cried. 'He frightened me. He never said anything at all. He just met me and we came here.' Parkhurst was now irreconcilably aware of the chemistry of her attraction for him and what it was that had touched him from the first moment of contact; something confirmed by those bemused, slightly too-large eyes, the placid calm that was almost—well, one could not apply the word bovine to such an exquisite creature—her silences, her surprise at deductions which must have been obvious, the naïveté of her revelations in reply to his questions. She was not only not quite bright, but irresistibly and abysmally stupid. Inside that sweet and alluring head nothing, but absolutely nothing stirred. And at that moment he felt himself swept by such a rush of love and protectiveness that it was hardly bearable. Later, of course, he was to realize that it was partly caused by the relief after years of having to live with his former wife Victoria's appalling cleverness. Such an endearing and wholly desirable creature housing such a vacuum was a prize beyond compare. But by then motivation no longer mattered; there was only Birdie.

He now paused to wonder about how far back the chain of command would reach. The hoodlum who had been sent to finger him would be, of course, the merest pawn. Was this merely an attack by the Dockerty–Schwab crowd, a crude attempt at infiltration, or did it extend all the way up to as high as one could go? The mysterious Uncle Nono? Parkhurst could afford to be patient. He had no doubt that in her good time this adorable creature would tell him that too.

He asked, 'Have you had any dinner? The *sauerbraten* with potato dumplings is wonderful.'

She said, 'Oh, I'd love it!'

She ate enjoying her food with the concentration of one whose mind is not taken up with anything else. When she had finished she wiped her lips daintily and gave him an angelic smile.

He decided to try shock tactics. He said, 'Uncle Nono sent you, didn't he?'

But this time the arrow seemed to have gone wild, for she regarded him with bewilderment, two furrows momentarily indenting the lovely sweep of her brow. Parkhurst added, 'You needn't be afraid, Birdie, it would have to come out anyway, wouldn't it?'

She asked, 'Who was it you said?'

'Well, he might have been called Uncle Nono, mightn't he?'

Birdie shook her head and said, 'But I don't know who you're talking about. I've never met anyone by that name.'

'Had you ever heard it before?'

Birdie lifted her head higher in an attempt to make trying to remember easier. Parkhurst was aware that thinking for her was a considerable effort. 'I don't think so,' she said and he felt that she was speaking the truth. Not a great deal penetrated to Birdie.

Parkhurst said, 'But somebody sent you to make contact—to meet me, I mean?'

Birdie turned her head away but the Duke caught the shine of a tear in the corner of one eye. She said, 'I'm so ashamed. But he was terribly nice. He was so awfully good to me when my mother was sick and needed an operation. I never asked. He sent her a thousand dollars. It saved her life, the doctor said.'

'Where did you meet him?'

'He came with some friends after the theater once, and we all went out to dinner. He said he loved my dancing. And afterwards we went out together two or three more times; that was when I told him all about my Mum.'

'Did he tell you his name?'

Birdie said, 'Well, just Joe. Everybody called him Joe, so I did, too. That's who he used to say he was when he called me on the telephone to ask me out.'

Parkhurst nodded and thought to himself so this, then,

was the pay-off for the obligation—go and pick up Duke Parkhurst; throw barbs into his heart. It was horrid and yet she was utterly irresistible.

Parkhurst asked, 'Could you describe him, Birdie? The man named Joe, I mean?'

She once more went through that excruciating agony which indicated she was trying to remember. She said, 'Well, he was dark and had sort of nice eyes . . . and . . .' and here her descriptive powers ran dry. Parkhurst thought that she probably would not even recognize the man if she were shown a picture of him.

She had a moment of illumination, 'But he was a gentleman.'

'How could you tell?'

'Well, you know, don't you? I suppose it's something you feel. You are.'

'Am I? What makes you think so?'

Birdie gave him a long, appraising look and her reply was the perfect one of innocence, 'I suppose because you're gentle,' she said, 'and clever.' Then she added, 'How did you know about the . . . Well, about the autograph book not being . . .'

'Birdie, Birdie!' Parkhurst explained, 'a girl with a family of a mother and two sisters in Milwaukee, and who had only been in New York two years would hardly be likely to have a fourteen-year-old brother in a New York high school, would she? You probably have no brother at all.'

Birdie only shook her head with the wonder of him.

'Besides which,' Parkhurst concluded, 'it only takes half an eye to see that the ink of those autographs of celebrities is quite old—some even turning rust-colored. The ink in which 'Bobby McBride aged 14. DeWitt Clinton High School" is written, is quite fresh; as fresh as my own signature.'

Expression at last came into Birdie's huge eyes, awe and marvel. She whispered, 'I think you're wonderful!'

It had grown late. Eventually they began putting out the lights in the main part of the restaurant and only a few stragglers were left sitting in the celebrity section. Parkhurst looked at his watch; it showed a quarter past one. He paid up and said to Birdie, 'Shall we go home?'

Birdie looked startled. 'Home with you?' she asked.

Parkhurst said, 'Isn't that what you're supposed to do?'

Birdie hung her head. 'I'd forgotten.'

Parkhurst asked, 'Would you rather not?'

Birdie was seized with that endearing embarrassment of a simple person who has suddenly discovered a desire. She whispered, 'I didn't know I'd be wanting to.'

'And supposing you hadn't wanted to?'

Birdie averted her eyes again and replied, 'He was so kind and generous to me when I was in trouble and never asked for anything back.'

'Until now?' Parkhurst said. He could only see the top of her head with the swatch of brown hair.

'Yes,' she whispered, 'until now.' And then so low that he could hardly hear it, she whispered, 'I'm not a nice girl.'

Parkhurst said, 'But you're a gentleman, Birdie.'

She looked up startled, 'A gentleman? I don't understand. What do you mean?'

He replied, 'A gentleman always honors his obligations. I'd like you to come home with me. Will you?'

Very softly, 'Yes, please.'

On the way out Parkhurst thought hard. The plot was what? Put a spy in his camp? Supply him with a mistress who would be manipulated by the enemy? The badger game? Blackmail? None of them made sense. He and Birdie were both free, white and over twenty-one. Jail bait, the age involving statutory rape, was sixteen. Besides which, he came back to his first conclusion, Birdie McBride was neither an unknown little tramp nor a call girl but someone who had begun to make her mark in show business and hence was known. Where was the catch? Long term? Short term?

And then he thought of something and it shook him for a moment that they would use this girl and why it did not matter to them that she was so stupid, but only that she was so damnably and irresistibly attractive. What had come into his mind was something that no one ever thought of any more, or mentioned or, it seemed, for years had got into the newspapers. There could be no question of any immediate danger. They had simply taken a long shot.

As the doorman called them a cab, Parkhurst did not feel himself under any observation. If he was right, none was necessary. He gave himself up to the present and to the emotions that had laid their hold upon him earlier in the evening. All the way up to his apartment on Park Avenue and 66th Street they did not speak to one another but Birdie clutched his arm and pressed her small face hard against his shoulder, and he held her hand.

Later they went to bed and their first experience together was successful and Parkhurst found it shatteringly beautiful.

The following morning Birdie was sitting up with the sheet over her, embracing her knees with a leftover smile of dreamy satisfaction touching her mouth. Her light brown hair tumbled down the long curve of her spine. Parkhurst was leaning on an elbow, satisfying his senses with looking at the girl but his thoughts kept on turning back to Matilda, the perfect fighting machine. He wondered whether Birdie would be flattered if he told her that she was the perfect loving machine, that never before had he known such a night, never had he felt so soothed, calm and happy.

Birdie asked, 'What are you thinking of?'

Parkhurst replied, 'You,' and then added, 'and Matilda, the boxing kangaroo. Have you ever heard of him?'

Birdie shook her head and said, 'What is a kangaroo like?'

'A marsupial indigenous to Australia. The female carries her young in a pouch just below her belly.' Parkhurst wished he had not used words like 'marsupial' and 'indigenous' to Birdie. Now that he had begun to understand her better it was somewhat like hitting a child.

But she only looked upon him with a return in her eyes of that awe, 'Oh, please tell me more about them. I love to hear you talk. We never had Australia back in Milwaukee. I was fourteen when I went to dancing school.'

Parkhurst expounded upon the marvels of Matilda until he realized that Birdie was not understanding a thing he was saying but was just enjoying the sound of the words and this somehow made him happier than he had been in a long time so that he leaned over and kissed the cool side of her temple saying, 'Oh Birdie, Birdie, you're wonderful!'

To his dismay and delight he realized that he was

helplessly in love with the little Broadway dancer who had not a brain cell working and who by and large was as ignorant of the simplest aspects of what went on around her as a two-year-old infant. At this point Birdie leaned over to him, put her face into the crevice of his shoulder and began to cry exactly like such a one.

At last what she was trying to say between sobs became intelligible, 'I'm such a horrid creature, Duke, and I know I'm so stupid. I don't know anything at all and I've fallen for you. I don't know what to do.'

Eventually she had done with crying and moved away a little from him, resuming the embrace of her knees and regarding him with a mixture of adoration and concern. Some thought, some speech was struggling inside her trying to take shape where it could be translated by her pretty mouth. Several times the lips parted and then closed again. Two furrows indented the lovely sweep of her brow and the fingers entwining her knees fluttered helplessly.

Parkhurst came to her rescue. 'It's about the man Joe, isn't it?' he said.

She gave a little chirrup of satisfaction and relief. 'Oh,' she cried, 'then you aren't angry with me.'

Parkhurst said, 'No, Birdie,' and then asked, 'What exactly did he ask you to do?'

She replied, 'Oh, nothing awful, just to be friends with you. But he didn't say to fall in lo . . .' And here she released a hand from her knee, clapped it over her mouth so that her voice was muffled. 'Oh, I said it, didn't I? I oughtn't to have.'

'Yes, you did, Birdie.' Parkhurst grinned suddenly and said, 'Beautiful spy falls for intended victim. Right out of Ian Fleming.'

The furrows appeared again in the brow, 'Ian . . . ?'

'Never mind,' Parkhurst said.

'But he didn't say anything about spying,' Birdie insisted. 'I'm not a spy. I wouldn't know how.'

Parkhurst knew that this was true enough. If it was Uncle Nono and he wanted inside information he would not send Birdie. No, she had told the truth; just a friendship; warm, beautiful, dangerous friendship with only one miscalculation.

Parkhurst said, 'I didn't mean you were a spy really, Birdie.' The girl was looking miserable again. He asked, 'Have you any idea who Joe really is?'

Birdie shook her head, 'No. Just that he was terribly nice. He was awfully good to me but I hadn't seen him for months until he called me and asked me to do him a favor.' At the word 'favor' she looked up at him quickly and her eyes were filled both with misery and terror. 'Oh, Duke,' she cried, 'will you hate me now that I've 'fessed?'

Parkhurst knew then in his heart that even such locutions as ''fessed' would be bearable with Birdie. Whatever the plot might be, the instrument of it that had been sent him was that most rare jewel, a true innocent and probably the one person in the world capable of forever closing over the deep wounds left by his marriage, memories of humiliations that had never healed. He was enraptured by her and was to become even more so. He smiled at her and said, 'Never! Not as long as you continue to carry out Joe's instructions so charmingly.'

She said, 'You mean be your friend?'

'Yes. Whatever the obligation, you may consider that you've paid off.'

She gazed at him uncomprehending and then shyly asked, 'Oh, Duke, could I come into your arms?' She did so.

He felt that she could be his heart's delight forever.

Two days later she moved in with him and there began for Duke Parkhurst the most contented period of his life he had ever known and only then did he realize how close his first marriage had come to destroying him. He could take this wonderful creature everywhere, in any company and she would sit and listen for hours to the talk, never interrupt or comment or offer an opinion, but only by her joyous beauty and irresistibleness help the talkers to expand. She was loyal, loving and faithful. For her, the sun rose and set with Duke Parkhurst. She never read a line he wrote, never criticized him, only loved him.

It was the same when he took her out of town on weekend trips. Everywhere she sat silently listening. She excited only admiration, friendship and good feeling.

It was on the train returning from their first journey away from New York together, a one-day affair, since Birdie was still in the show then and had to be back for her performance that evening, that Parkhurst asked her, 'By the way, Birdie, are you being paid for this?'

They were in a day coach and she was nestling to him in the seat even more closely, pressing her cheek so deliciously hard against his shoulder, her eyes shut, Now they opened with a startled expression and she nodded guiltily.

'How much?'

'Two hundred dollars a week.'

'As long as we remain friends?'

Birdie nodded again and added, 'I'm so ashamed, Duke. Especially now, after...'

Parkhurst had burst into roars of laughter and had continued laughing for the rest of the way. For he had said to her, 'Take the money and keep it for a little nest-egg, as long as you can. It's his bargain, you know.' It seemed only justice to him under the circumstances that she should do so, as long as she was delivering the goods.

# XIV

THE FORCES OF EVIL were not yet done with their attempts to
bring about the retirement of Matilda and company and halt
the triumphal swathe he was cutting through the middleweight crop, though since Duke Parkhurst had recounted
the merry affair of the three hoods and the sheriff at Clayton,
Oklahoma, there were no further attempts at murder most
foul. The organization never made the same mistake twice.

Nevertheless, now alerted, Patrick Aloysius arranged for
maximum security as far as they were able to do so. A
twenty-four-hour watch was instituted over Matilda; Billy
Baker slept with him at night in his stall armed with a cannon
of more manageable size than Old Smokey, a parting gift
from Sheriff Sam Baker, and particular attention was paid
to the fruits and vegetables as well as goodies fed to Matilda
to make certain that nothing lethal was introduced.

But as Patrick Aloysius well knew, and Bimmie was
learning, there were other means of embarrassment available
and unfortunately there was no way of anticipating when
and where they might be met and countered.

Thus in the Auditorium in Eagle City, Nevada, a famous
gaming center second only to Las Vegas, where Matilda was
engaging one Soldier Barton he found himself in the ring
with a referee suffering from myopia, general forgetfulness,
two left feet and what appeared to be indications of a long-
standing friendship with, and affection for, the Soldier.

From the outset of the contest he kept getting in the way
every time Matilda was poised for a clean shot at the
Soldier's prominent chin. He broke up the rhythm of his

attack by walking between him and his opponent at every opportunity. He let Barton hang on to his heart's content in the clinches and indulged in every known trick by which a referee who has been suborned can make life difficult without apparently doing anything that might look illegal or prejudicial to the spectators.

Old fox Ahearn spotted the symptoms immediately, as did Billy Baker and began screaming at him until they were warned to shut up by a ringside policeman on pain of being thrown out.

Matilda got in one swipe which draped the Soldier over one of the lower ropes and retired, as always, to a neutral corner as the knock-down timekeeper was tolling off the seconds. The referee, however, assisted the Soldier to his feet, ministering to him tenderly, brushing off his gloves, giving him an apparent examination for injuries and all in all securing him thirty seconds of respite while the triumvirate raged helplessly at the foot of the steps in their corner.

Ahearn said, 'The son-of-a-bitch is bought!'

Billy Baker said, 'I'm going to throw in the towel! The bloody barstid will get Matilda 'urt!'

Patrick Aloysius snarled, 'Throw in the towel, when Matilda's winning? They'll tear the place down!'

Bimmie cried, 'Ain't there nothing we can do?'

There was not, and Patrick Aloysius knew it. During the three minutes of combat in each round the referee is in absolute control and no one can interfere. Afterwards, if his work has been too blatantly crude, there can be an investigation but by then it is usually too late and the damage has been done.

Bimmie cried, 'We got to do something!'

The bell rang. Patrick Aloysius said, 'Let him go another round and if it continues, we won't let him come out for the third, take the technical kayo and put in a protest.'

During the minute's rest Baker had seized Matilda's left ear and, using it as though it were a microphone, was pouring instructions into it. 'You've got ter watch yerself, chum. That bloody referee's out to do you in. He'll try to step on your feet or hold yer arms. You've got ter keep away from 'im. Don't let him come near you. I had one like that once in a match in Melbourne; it's like trying to fight two blokes at the same time. You've got to 'ave eyes in the back of yer 'ead. Keep away from both of them. Box one and don't let the other come close. There ain't no use zapping the Soldier, you saw what 'e did the last time you dropped 'im. Keep circling and don't let 'im go into a clinch. You'll be all right as long as you don't let the bloody referee have a chance to get near you.'

Patrick Aloysius, who was feeling sick at heart, said wearily, 'Any other instructions? I thought you said he couldn't understand a word anyone said to him?'

'Nor he can't,' Baker replied, 'but how can you be sure? What if he could? I wouldn't be doin' me duty if I didn't give him the benefit of me experience. He ain't never had anything like this 'appen to 'im before.'

The klaxon sounded for seconds to leave the ring. As he went down the stairs, Baker called back, 'Remember what I told you, Matilda!' The bell rang.

Matilda was out of his corner with his usual mid-ring hop to be met by both the Soldier and the referee. It was obvious that he had comprehended nothing of Baker's warnings, for with a slightly bewildered furrow on his brow he let Barton go into an immediate clinch, with the referee pulling and hauling at him so that for a moment there seemed to be a triple embrace going on.

'Oh, my Gawd!' moaned Billy Baker. 'He's goin' to zap the referee! He's given 'im the kiss of death! He'll be disqualerfied.'

It was true. For Matilda's heart was suddenly overflowing with love and happiness at the new game with which he had been presented—two for the price of one, so to speak—and all the unusual intelligence which had become centered in ring craft was called into play to cope with the situation. He had worked it out much as he would have in the bush of his native land had he found himself under simultaneous attack by two other kangaroos.

On the next time around, as the referee approached him, Matilda swivelled on his haunches and catching the representative of the Forces of Evil in the midriff with his powerful tail, hoisted him out of the ring, depositing him in the lap of a wealthy gaming house owner sitting just beyond the Working Press. Using the backlash of the extra impetus gained from this manoeuvre, Matilda uppercut the Soldier clean over the top rope and into the bosom of a lady friend of a prominent cattleman in Row C, for what proved to be a new distance record when measured later. By the time both the Soldier and the referee had managed to regain the platform the timekeeper, who someone apparently had carelessly neglected to suborn, had counted up to thirty and it was all over.

Everyone in the auditorium felt that the ending was the most rewarding and satisfactory they had ever witnessed in a boxing match. Billy Baker was so ecstatic he rewarded Matilda with three Hershey and two Mars Bars. Even Bimmie embraced the animal, leaving only Patrick Aloysius to interest himself in the jawing match going on in mid-ring, centered around the accusation made by several citizens no one seemed to know, that Matilda should be disqualified for striking the referee during the course of a bout. Patrick Aloysius was at pains to point out that a tail was not a striking weapon, that the referee should not have been allowed to officiate at a struggle between a couple of two-year-olds for the possession of a lollipop, that he could not get out of his own way and that Matilda had merely been trying to avoid the onslaught of his opponent when his appendage had pipped the arbiter.

Thereafter the authorities ruled that the bout had been well and truly fought, that Matilda was the winner by a knock-out after fifty-seven seconds of the second round. The Forces of Evil retreated like demons at cock-crow, muttering imprecations.

The record book shows that Matilda disposed of Mickey Portal in thirty-seven seconds in Paint City, Utah; Patsy Hogan in two, fifty-three in Springville, Idaho; and Frankie Delane at Carson Forks, Oregon, in two, thirteen; all in the first round, all good journeyman middleweights. These victories were beginning to attract considerable attention in the local press and some of the national newspapers as well and even the New York dailies, rivals of the *Mercury,* were listing the affairs briefly in six-point type, under the generic heading of 'OUT OF TOWN RESULTS', and would report them as, 'Matilda, 159½, knocked out Johnny Clabby, 160, at Three Rivers, Washington, last night, putting him down for the full count with a left to the body in two minutes and twelve seconds of the first round', conveniently neglecting to add that Matilda was a kangaroo. But by this time practically everyone in the United States knew and if they did not, it was not the fault of Duke Parkhurst.

Via his inside rollicking stories an ever increasing portion of American boxing fans were being brainwashed into thinking of Matilda as the middleweight champion of the world, with Dockerty and Company playing further into his hands by keeping out of sight and out of the news, as ordered.

This series of triumphs was advancing Matilda's cause and led to a more important match with a rising Western contender, Rocky Ruffalo, whom they engaged to meet over the ten-round route in San Martino, California. Here the Forces of Evil struck again.

It was not that Billy Baker had become neglectful or that the triumvirate had allowed themselves to be lulled into a sense of false security, as several months had gone by since the affair of the purchased referee, but only that San Martino is one of those bright, neon-lighted Californian coast towns, bustling with business and traffic and it seemed nothing untoward was likely to happen here.

They were due at the arena at nine o'clock to be ready for the ten-o'clock main event. At a quarter to nine, Billy Baker said to Matilda, 'Blimey, old cock, I'm out of gaspers. I won't be a minute.'

The tobacconist's was only a half block down from the shed which had been fixed up for Matilda's convenience. Car traffic was heavy. There were many people in the street, going and coming. The neon tube display signs of cinemas, supermarkets and stores, were turning the night into multicolored day. Secure at being surrounded by normal, busy humanity, Baker made his way to the shop.

The affair of his removal was so beautifully and efficiently handled that even he was unable to withhold grudging admiration. As he approached the tobacconist he passed a

143

snappy-looking yellow and white sedan; two men in casual clothes and a girl in an attractive floral cotton frock were standing at its side chatting. As Baker drew nigh, all three suddenly turned to stare at him for a moment and one of the men said, 'Hey, hello Jack! When did you get into town?'

'Jack, darling!' the girl squealed and threw her arms around his neck, kissing him affectionately.

For just that fraction of a second too long Baker was under the impression that it was a genuine case of mistaken identity until the second man had closed in behind him and was relieving him of the pistol in his hip pocket and at the same time pressing something hard against his spine, saying softly into Baker's ear, 'Keep smiling, pal. You're just as glad to see us as we are to see you, ain't you? Now we'll all just go for a nice little ride together and nobody's going to get hurt.'

It seemed impossible on this busy street with people on their way to the movies, a policeman directing traffic on the far corner, cars passing every second. The man behind him felt Baker's arm muscle tense and read his mind that he was going to try a break-away. The other two were still enthusing over the chance meeting with their old friend Jack. The voice behind became chill, 'Don't do it, pal, or you'll be entered as D.O.A. when the meat wagon delivers you to the hospital. Just get in the back like we're all chummy.'

Billy Baker wanted to stay alive as much as anybody else. He got in. The fellow with the gun sat beside him. The girl and the other man who drove, occupied the front seat. There was no attempt at hurry, or even to go anywhere in particular. To all intents and purposes they were merely out for an evening's drive and when they passed the cop on point duty, the girl gave him a friendly smile and a wave.

At least, Baker felt, they did not seem to be on the point of killing him at once. He said, 'What's the idea? What the 'ell do you bleedin' sods want of me?'

His captor at his side replied, 'Just your company, friend, just your company for a little while.'

Billy Baker was beginning to have an inkling and now was regretting that he had not taken a stab at escape. He said, 'What if I hadn't got in and had yelled, or something. What was that D.O.A. business at the 'orspital?'

His seat mate replied succinctly, 'Dead on arrival.'

Baker remembered that now and it silenced him for a time. They would have done it, too, and got into their car and driven off and been lost in the shuffle. He had one more question, for he was curious as to how good the organization might be and whether there was any chance. He said, 'What would you 'ave done if I 'adn't gone out for some fags?'

'You'd have been picked up at the arena,' said his friend, 'maybe when you went to the john.'

The girl giggled. They drove a block off the main highway and parked in a side street with the windows rolled up. At ten o'clock promptly the girl in the front seat reached out and turned on the radio, a gesture that was matched in homes and cars all over town. All those who had not been able or did not care to go to the arena, were most certainly going to get the fight that way. The torture of Billy Baker being compelled to listen to the débâcle was cruel, unusual and gratuitous.

The announcer, being a local boy on the local station, was inclined to be voluble and over excited but withal no bad hand at furnishing a picture of what was going on.

'... Yes, here they are. I can see them now coming down the aisle, Matilda hopping along. There's Paddy Ahearn with him and Bimmie Bimstein, his managers. I don't see Billy Baker yet, but I guess he'll be coming along behind. Here they come. Rocky's already in the ring and dancing around ... Wait a minute, there's some sort of trouble. Matilda doesn't seem to want to go any further. Ahearn is pulling on the chain attached to his collar. Matilda is standing up now, on his hindlegs. He seems to be looking around for something or someone.'

'Cor,' said Billy Baker, 'the poor barstid! It's me 'e's looking for.'

The man in the back next him said, 'Well, waddyer know!'

'... They've got him moving again now, just the two of them. There's no sign of Billy Baker yet. There's something funny going on, but I don't know what it is. Matilda looks nervous. It's all Ahearn and Bimstein can do to persuade him to hop into the ring. Now he's standing up again and looking around ... They're having trouble putting the gloves on him.

The two managers keep stroking and trying to soothe him. He's all worked up. Maybe it's because Baker isn't with him. Wait a minute, I'll ask... I guess that's it—but Paddy Aloysius Ahearn his manager said he'll be all right in a minute. He's just a little upset. His trainer Baker was taken sick and couldn't come.'

'Taken sick, my arse!' said Billy Baker. 'You bloody barstids, 'e won't fight without I'm there!'

The man at the wheel turned around and spoke for the first time. He said, 'Now ain't that too bad!'

The broadcaster pitched his voice a half tone higher and resumed his description, 'They've got the gloves on him now and the referee is calling them to the center of the ring for their instructions. That's funny, Matilda won't shake hands. He's acting up and they're trying to make him. He usually gives his opponent a kiss at this point but he hasn't. He's sort of frantic...'

'Cor blimey! You bloody swine! Lemme go!' pleaded Billy Baker.

'Shut up!' said his seat mate and then to the girl. 'Turn it up louder.'

'They're back in their corners now. There goes the bell! Rocky advances to the center of the ring but Matilda is just standing in his corner. He's not coming out! He isn't doing anything. He's just standing there looking around. Rocky comes over but Matilda has his hand down and Rocky doesn't know what to do. Maybe Matilda hasn't heard the bell. It looks like Matilda just isn't going to fight!'

Billy Baker moaned, 'I knew it! I knew it!'

The man in the front seat said, 'Goodbye, Matilda!' and the girl giggled again.

'... Now the referee is motioning for Rocky to go in and hit Matilda, but Matilda suddenly moves away along the ropes. He's looking out into the audience... He's looking worried. He's hopping along the ropes. I guess maybe he's looking for Billy Baker.

'Poor Rocky still doesn't know what to do, he's just being ignored. Matilda has his back to him. Rocky goes after him with his right cocked, but Matilda simply hops away from him. He's hopping all around the ring, looking from each side. Rocky is bewildered . . . The crowd is beginning to get restless and boo and handclap—you can hear them! Rocky must be wondering whether he'd eaten garlic or something. In Matilda's corner Ahearn and Bimmie are making motions for him to go in and fight. They're gesturing with left and right hooks, but he's ignoring them too. Now he's down on all fours peering into the Working Press.'

Billy Baker groaned, 'I can't stand it!' and tried to cover his ears with his hands.

He was warned, 'Don't move, pal! It won't be long now.'

'. . . Nobody knows what to do. Matilda is on all fours and you can't hit a kangaroo when he's down. No blow has been struck so it isn't a knock-down for a count . . . You can hear the crowd getting sore. Now Matilda's standing up again, hopping away and looking up into the balconies. Oh, oh, the referee goes over to him, takes Matilda by the shoulders, whirls him around and motions him to get in and fight. Matilda pays no attention. He's snubbing the referee just like he has Rocky. He's now gone to the other side of the ring. The referee is going over. I think he's going to disqualify Matilda. Yes, he's going to disqualify him! Oh, oh!' And here the broadcaster's voice suddenly rose into a hysterical scream, 'Wait a minute! Rocky's down! Rocky got in Matilda's way as he was moving to the other side to look up into the center balcony, and Rocky's down. Matilda has just hit him! But he didn't seem to mean to. It was just because he was in his way. It was a right—a right to the jaw! Matilda hit Rocky with a right to the jaw! Listen to the crowd yell! The referee's counting. Oh boy, what a punch! Yup, that's it—eight . . . nine . . . ten . . . and out! Yes, he's out! He never moved! Maybe he wasn't looking for Baker at all and it was a trap. Rocky walked into the trap and Matilda caught him with a perfect right. Rocky hasn't stirred yet! Can you hear the crowd cheering for Matilda?'

Yes, they could indeed hear the sound of the cheering for it now drowned out everything else, the clanging of the bell, the announcer giving the time, everything except Billy Baker's yell of delight, 'He shouldn't 'ave got in Matilda's way! He gets narked when he's crossed.'

The man at his side had the gun drawn back to pistol whip

the little Cockney. It was the girl who interfered. She said, 'Don't! He hasn't done anything. We're going to have enough to explain. I wanna get out of here!'

The man opened the door and pushed Baker into the street where he fell on all fours. Dust and dirt whirled into his face as, from a standing start, the car roared away and around the corner with a screech of tires. He picked himself up unhurt, found himself only three short blocks from the arena and started to run.

Such had been the excitement over Matilda's strange tactics and the sudden unexpected knock-out, plus the fact that it took longer than usual to revive the unfortunate Rocky, that Matilda, Ahearn and Bimmie had not left the arena when the dishevelled figure of The Bermondsey Kid was observed running up the aisle from the entrance, to leap into the ring where the most touching scene took place between him and Matilda. The animal gathered him in its embrace and covered his face with affectionate kisses while happy little noises which those in the Working Press described as, 'Uck-uck-uck-uck-uck', broke from his throat. And while no one in the audience actually knew what had happened the reunion and Matilda's joy was so affecting that many in the lingering rows of spectators found themselves moved to the shedding of unashamed tears.

Once more a bout involving Matilda had provided a sensational ending and the local press gave it red hot coverage; the national dailies began to sit up and take notice. But only Duke Parkhurst had the inside story of the kidnapping of Billy Baker, as relayed by Bimmie over the phone, for his late editions.

The three, of course, were never apprehended, the white and yellow sedan having been found abandoned. Nor did any extraordinary corpses turn up anywhere which might have been connected with the fiasco. The fact that Rocky Ruffalo had momentarily obstructed Matilda's search and got himself zapped in a rather absent-minded way could only be set down to bad luck. Besides which, the fact that Birdie had given up her dancing and was seen with Parkhurst where ever he went was a source of growing satisfaction to the Forces of Evil. If other ploys were going awry, Plan B was not. The columnist was sticking his head further and further into the noose and Uncle Nono need only bide the time and the place to draw it tight. In the meantime more schemes for upsetting the upward climb of the triumvirate were being set afoot.

# XV

TWO MONTHS LATER, as the leaves were beginning to turn, Matilda and Company were back east where the next build-up fight was scheduled. Parkhurst was closeted with Justus Clay, his managing editor. Bimmie, in a state of euphoria, was on the subway, northbound, to see his girl Hannah.

Parkhurst and Clay were old friends and drinking companions off the premises of the *Mercury* but within, the distinction between managing editor and sports editor remained. Parkhurst with his sports column was the star turn of the newspaper and, as has been indicated, a potent New York character, but it was Clay in the final analysis who was responsible for everything that went into the paper.

Clay was saying, 'It's been great, Duke. And a big laugh. But how long are you going to keep it up before people start laughing at us, instead of with us? The Commander loves a good joke, as you know, but not when it's on him or his paper.'

The Commander was Jason Hathaway, publisher of the *Mercury,* but around the office known only by the rank he had held in the Navy during the war.

Parkhurst said, 'It's building circulation and will keep on doing so until those chumps on the other papers have the sense to pick it up. We've put on close to three hundred thousand.'

'I know,' Clay said, 'I'm having it analysed.'

Parkhurst said, 'We've got the Hippies with us.'

Clay's permanently sour expression showed momentary surprise, 'What?' he said. 'How do you know?'

Parkhurst grinned at him, 'They've adopted Matilda. He satisfies their sense of anarchy; the middleweight championship of the world held by a kangaroo. There's revolution for you. They've been thronging his fights out West.'

He reached into his pocket and produced a handful of celluloid badges, replicas of Matilda in boxing position with slogans superimposed, 'MATILDA IS LOVE'; 'OUR CHAMPION MATILDA'; 'LOVE THAT KANGAROO'; 'KISS FIRST THEN ZAP'; 'SEND MATILDA TO VIETNAM'; 'IF YOU MUST FIGHT, LET MATILDA DO IT'. Parkhurst said, 'There's a new pop song coming out, "Kiss of Death". It's all about the last rites administered by Matilda before he lets go.'

Clay examined the badges, pulling at his lower lip reflectively. He said, 'Hippies don't buy. You won't impress the advertising department with that.'

Parkhurst said, 'Since when? Any bunch that can travel by car a hundred and fifty thousand strong to a mountainside in Vermont, to listen to pop singers, has purchasing power. They can buy records if nothing else. What are you going to do—let down the youth of the country? Matilda's their new hero.'

Clay's expression had returned to normal disapproval. He said, 'Well, if you think you know what you're doing...'

Parkhurst said, 'There's something else, Justus. But it's still pretty dicey. I think it's beginning to get under Uncle Nono's skin. If we could smoke him out...'

Clay perked up again, 'What do you mean—identify him?'

Parkhurst said, 'He's beginning to show signs of irritation. Everybody knows that Pinky Schwab is just the manager of convenience for Lee Dockerty and that his contract is really owned by a hood named Marcanti. But who's behind Marcanti? For want of identification he's known along the Stem as Uncle Nono. The point is that Marcanti isn't bright enough or powerful enough to have engineered some of the things that have been happening to these Matilda people. *Ergo,* Uncle Nono getting edgy. If I keep on sooner or later he might go just that much too far...'

Clay asked, 'Have you any idea who this Uncle Nono might be?'

Parkhurst replied, 'A guy named Joe,' and then wished he hadn't.

Clay said disgustedly, 'Brother, you're a help!' And appeared to accept it as a gag, when he suddenly regarded Parkhurst shrewdly and asked, 'You wouldn't be having any ideas or inside information on this Uncle Nono that you've been keeping from your Uncle Clay, would you? Joe Who?'

Parkhurst said sharply, 'Joe nobody!' He felt like a horse in mid-race that has been reined in sharply. Joe Who, Birdie's friend, who had been kind to her, who never gave her a last name or an address or a telephone number, who only called her at the theater; who, if he had stayed with her probably had done so at her flat? Joe Nono. Joe Nothing. But what if Parkhurst were to succeed in identifying him and he held Birdie responsible? The track was no longer clear and fast. The going might become muddy.

Clay said, 'No offense, no offense, but it might be nice if we could...'

Parkhurst found himself talking in terms of the image that had formed in his head of being pulled up. He said, 'I'd just as soon lay back of the Feds on this for the time being and let them make the running. If we do tease Uncle Nono into a mistake, we'll have the inside track with the F.B.I.'

Clay said, 'I suppose you're right, but you'd get a Pulitzer Prize if you cracked it.'

*And a coffin for Birdie,* Parkhurst thought.

'Okay, then, Duke go ahead,' Clay concluded, adding, 'But don't stick your neck out too far. By the way, just how far are you planning to carry it?'

Parkhurst had already started for the door and threw over his shoulder, 'All the way.'

Clay looked up in alarm and yelled, 'Come back here, you! What the hell do you mean by all the way?'

Parkhurst sauntered back slowly, his hands in his pockets. Well, it had to come out someday, the big dream, and now was as good a time as any. He knew that Justus had been impressed with the circulation figures. 'The rematch between Lee Dockerty and Matilda at Jericho Stadium for our Free Food Fund for Hungry Children.'

A stricken cry emerged from Clay. 'What? Are you kidding?'

Parkhurst kept his huge bulk moving towards Clay's desk and towered over him to say in imitation Broadwayese, 'Ain't you heard that Duke Parkhurst, he don't kid?'

Justus Clay looked up at him, his monkey face contorted in genuine agony, 'You're crazy, Duke! Since when have you been on grass? You're either off your nut or fossilized!'

'No, I'm not,' Parkhurst said, 'we could gross three million dollars. I could turn over a million and a half to our Fund, maybe more.'

'A man against a kangaroo here in New York? You'd never pull it off in a million years!'

Parkhurst said, 'If Matilda can get away with his next two bouts—the one next month in Philadelphia and then Chicago—Pinky and Dockerty have got to give him a return match. If they don't, I'll hound them both out of the country.' And then almost casually he added, 'I've already had a talk with Wally Mason, President of the Stadium Corporation. He neither laughed at me nor screamed like you.'

Clay leaned back in his chair and looked up at the big man, and said, 'You're serious, aren't you?'

'Dead.'

'How much did you say?'

'Three million gross. A million and a half net for the charity. The Commander might like that.'

Clay said, 'You'd never get a licence. Wild Bill wouldn't dare.'

'We might be able to persuade him,' Parkhurst replied. Then subtly including Clay, 'If we do, will you back me?'

Clay was silent for a moment and then began to doodle on a scratch pad. 'A million and a half,' he said. 'Christ!'

Parkhurst knew he was hooked.

Bimmie and Hannah sat side by side on the couch alone in the living room of the Lebensraums' duplex apartment at 87th Street and West End Avenue. On Hannah's left hand sparkled a square-cut diamond solitaire and there was

153

nothing cheap about it, even though of course Bimmie had got it at wholesale price from a friend in the diamond trade. Seven grand, he had paid for it and Hannah's bosom was subjected alternatively to thrill as she looked at the beautiful stone, and butterflies as she contemplated how Bimmie had managed to lay his hands on that much money.

Bimmie's wavy blond hair had been slicked back and was content to rest in aromatic layers. He was barbered and clad in his first really sharp suit of clothes with all the accessories.

Earlier at dinner, he had been exposed to the scrutiny of the Lebensraums: Mr and Mrs and Hannah's kid sister Myra, aged nine. Myra was nursing the biggest doll she had ever seen, as tall almost as herself. Generosity to others as well as himself had always been Bimmie's nature, and Henry and Sophie Lebensraum were discussing him now in the kitchen as a prospective son-in-law, rather than someone who needed to be fended off or made to feel unwelcome.

'He's a nice boy,' said Mrs Lebensraum, 'but so quick he made all that money? I can't understand it. Only six months ago Hannah was asking me could she borrow five dollars so she could take him out to dinner.'

Mr Lebensraum said, 'It only goes to show how clever he is. Anyone who can come up from nothing who is even something of a schlemiel and can suddenly buy diamond rings, presents for Myra and sending flowers, is smart enough to take care of our Hannah.'

'I wouldn't argue with you, Henry,' said Mrs Lebensraum, 'he's got nice eyes and the way he looks at our Hannah, he'll be good to her. But I don't understand how he made it so quick, and what kind of business can you get rich so fast, unless maybe there's something...'

'Didn't you read and see his picture in the paper the other day?' said Mr Lebensraum. 'He's manager of Matilda the middleweight champion of the world, a box fighter.'

'So?' Sophie replied, 'I didn't see. Is the fighter a nice fellow?'

'Well, he's a kangaroo.'

'A what?'

'Well, an animal—a kangaroo. You know, from Austral-

154

ia. I've been reading all about him for months in Duke Parkhurst's column. Only I never knew that Hannah's boyfriend was connected with him. Mr Parkhurst says that he's the best middleweight box fighter in the world. Bimmie told me he's going to make a million dollars with him.'

Mrs Lebensraum shook her head marveling and said, 'My, my! Our little Hannah is going to be a millionairess.' And then regarding her husband anxiously, she asked, 'But how do I say to my friends when they are asking about our Hannah's husband? He's managing an animal from Australia?'

Mr Lebensraum looked annoyed and said, 'I told you, Sophie, he's a champion boxer as well as being an animal.'

Mrs Lebensraum said, 'I guess I just don't understand such things. But Bimmie's a nice, polite boy.'

In the living-room which the Lebensraums now studiously avoided in order to 'let the young people be by themselves', with Myra too being forbidden the premises, Hannah was also not understanding certain things and doubts and questions similar to those that had assailed her mother were disturbing her, too. The butterflies were beginning to win out over the thrills. She twisted the ring on her finger, half-way down the digit.

Bimmie asked, 'Do you like it, Hannah?' And then he added, 'There's plenty more where that came from.'

Hannah now slipped the circlet wholly from her hand causing it to give forth a flash of blue-white fire.

'Oh, Bimmie!' Hannah cried, with a curious kind of sob in her throat, 'I love it! It's beautiful! It's wonderful! But, Bimmie, please don't be angry with me over what I'm about to say. I know you won't be but maybe you will, and I must say it.'

Bimmie said, 'Sure, sure, baby. Anything you like. There shouldn't be anything you couldn't say between an engaged couple, should there?'

Hannah regarded the object sparkling in the palm of her hand with wistful longing for an instant and then blurted out, 'Oh Bimmie, is it honest?'

Bimmie stalled for time. He said, 'Is what honest?'

155

'Well, Bimmie, your being able to give me such a beautiful engagement present. Poppa said it must be worth almost eight thousand dollars.'

'Seven,' Bimmie interjected. 'I got a thousand off.'

'Well, but don't you see?' Hannah tried to explain. 'From being in a hole-in-the-wall with no business coming in and eating chop suey lunches because you're broke, suddenly you're in a new office with that man who spits tobacco and can buy eight-thousand-dollar rings.'

'Seven thousand.'

'Bimmie, you know we've been going together for two years now, and I wouldn't go with anybody else, but Bimmie, you would break my heart if anything you were doing wasn't honest. It would be the one thing I couldn't bear because on one side I'm so proud of you for being so clever and winning over Momma and Poppa—you should have heard the way they *used* to talk about you, but on the other hand there're so many things I don't understand: how a wild animal like a kangaroo from Australia should be a boxing champion. And all the money you suddenly have and talking about making a million dollars. I want to be able to be proud of you all the way, all the time.'

Bimmie remained silently contemplative for one more instant, summoning up his forces and then asked, 'What's honest today, baby? What does it mean?'

Hannah looked startled. She had expected, needed and wanted only a little reassurance and she felt her heart sinking.

'Why, everyone!' she exclaimed. 'People, business—I mean, I'm honest; at least I try to be. Poppa's honest.'

Bimmie threw a shattering question at her, 'You call a forty percent mark-up honest?'

Hannah looked even more startled, 'I don't know what you mean, Bimmie.'

He said, 'Don't get me wrong, baby. You know I'm for you and I got the greatest respect for Mr Lebensraum who is a wonderful man I'll be proud to have as a father-in-law, and also your mother. But what does he do? He buys an overcoat from the wholesaler for eighty dollars and sells it for a hundred and twenty.'

Hannah blinked, 'But that's business,' she said.

'It's business,' Bimmie said, 'but is it honest? What has Mr Lebensraum done for that overcoat besides his overhead that it suddenly should cost forty percent more? Do you know what Valentine's Jewelry Store on Fifth Avenue would charge you for that diamond? Ten grand. Just for putting a guy in a monkey suit to open the door for you when you go in. What kind of honesty is that? And the ritzier the joint, the bigger the robbery. But at least your old man and them other big stores don't tell you no lies about their goods.'

Hannah blinked again, this time to hold back tears gathering at the corners of her worried eyes. A state of confusion and with it more fear was settling upon her. 'Lies? What kind of lies?'

'Selling lies,' Bimmie said. 'Open any magazine and read the advertisements and see what they say, or listen to the commercials. Somewhere there's always a con. Some cream is going to make your skin look ten years younger; a new style in brassières is going to make your boy-friend fall all over himself. Some company's tires can give you extra mileage. What are you going to do, measure? A cigarette is smooth—what's smooth? Cars, airlines, perfumes, liquor— they all snow you. Give a little, take a little, stretch it here, exaggerate there, put up an argument you can't prove. You got to wake up a little, baby, nothing is a hundred percent on the level. This is a screwy world we live in from the top down, and I mean from the top, where the President is allowed to tell you things that ain't true, and every time you open the papers some Judge, or General or somebody who's supposed to be a big shot, gets caught with his hand in the till. Who says what's honest? Somebody. Then when somebody else says what was honest yesterday ain't honest tomorrow, they pass a law and you're a crook. Why do you think they got to have so many laws? Because otherwise there'd be more crooks than they got already.'

'Oh, Bimmie,' Hannah wailed, 'and I thought you were honest!' And now tears fell.

'Hannah, baby! Sugar! What do you want from me? I ain't never robbed any widows or orphans or got my fingers stuck in the collection box and my kanagaroo is a hundred percent on the level. He don't know no different. Animals ain't like people. They wouldn't know how to chisel. What I done, Hannah? I got me an honest kanagaroo who wants to be a world's boxing champion. What's wrong with that? Do you know how many kanagaroos there are boxing in Austeralia? There's million of them. One of them's got to be the best and that's my Matilda.'

Hannah was watching Bimmie doubtfully. He might have got away with it if he had not made that crack about her father's mark-up as being dishonest. Bimmie had hurt her. If he thought that the legitimate business of buying cheap and selling dear, or cornering a market or advertising your goods was crooked, goodness knows what kind of shady deals he himself was involved in with those two horrid men who were his partners suddenly—the one who was always chewing tobacco and the other who looked like a Pekinese with his pushed-in nose. Hannah had led a sheltered life, but she was a New Yorker and a college girl and by osmosis she had acquired the knowledge that the prize-fight business as conducted there and elsewhere was not exactly admirable.

The ring held between Hannah's fingers caught the light from the fringed crimson lampshade and to her suddenly seemed to reflect a red evil glow. She turned dark, pleading eyes upon Bimmie and asked, 'Bimmie, will you swear that everything you and those other men have done with that poor—like you say—innocent animal is a hundred percent honest? I've got to ask it of you, Bimmie. You wouldn't lie to me.'

Bimmie ran the question through the computer of his agile mind. He took the ring from her fingers for a moment to inspect it and give the electronic circuits time to let the digits fall into place. He had to consider his partners as well.

'Hannah, sweetheart, every business ain't the same and you got to go along like it says when you're in Rome, you got to slice the baloney maybe a little thinner. I love that kanagaroo like he was my own brother, even better and I got to look after him so nothing should happen to him. Like down south, where three guys were going to shoot him if we don't take a dive, which I swear on my grandmother's grave Matilda wouldn't do. Look Hannah, baby, put the ring back on and don't worry your head about things you couldn't understand. No matter what you hear, notwithstanding to the contrary, Matilda is a hundred and ten percent on the up and up. I swear it. When you go in for the champeenship that's when the monkey business got to stop. Is that all right now?' And he concluded once more with his theme, 'So what's honest today? You tell me.'

Hannah Lebensraum shook her lovely head sadly and there were tears falling as she penetrated for once through the fog of schmoos and said, 'Bimmie, you haven't answered my question.' And then she added, 'Yes, I think perhaps you have.' She pushed the ring away from her and said, 'Keep it, Bimmie. I don't want it.'

Bimmie's layers of hair went straight up in pure panic. 'You mean we ain't engaged any more? Hannah, baby, I can explain everything...'

The girl shook her head. 'You had your chance, Bimmie. There will be only more schmoos. You could talk anybody around, Bimmie, but not this time. I thank you for the honor of asking me and buying me such a beautiful and expensive ring, but I always told you, Bimmie, the one thing I couldn't stand is if something isn't honest.'

Bimmie was genuinely shaken. How could a life lived according to the rules laid down by the enclave bounded by 39th to 50th Street north and south, and Sixth and Eighth Avenues east and west, suddenly shatter upon an unforeseen rock of Hannah's ethics? He looked suddenly quite small, forlorn and lost as he asked, 'Don't it make any difference how I feel about you, Hannah? What I'm feeling inside from the thought of losing you? Everything I done, I done to get rich for you...'

'It makes a terrible difference,' Hannah cried. 'I feel the same way, but I can't help it. I couldn't marry a crook. Please, Bimmie, go away. I don't want to see you again.' She got up and ran from the room.

Bimmie looked into the varying fires of the solitaire and wondered whether he would cry himself. But he shook it off. In the world of big shots one had to be a man. He pocketed the ring, picked up his hat and left the duplex of the Lebensraums—forever, it seemed to him certain.

159

# XVI

EVERYTHING WAS GOING swimmingly when Patrick Aloysius was called away to the telephone, leaving Bimmie leaning on the ropes of the Camden Gymnasium, watching Matilda work out with Billy Baker for his match against Mickey Young three nights later in the Downtown Arena.

As befitted the manager of an important fighter on active campaign, Bimmie was now clad in a white turtle-necked sweater with blue edging and pockets, and 'MATILDA' in red letters with the Australian flag across the back.

Matilda actually did not need to train but the amazing phenomenon of a kangaroo who could box like a man drew the curious and at a dollar a head there were some three hundred spectators crowded into the gymnasium. The money was rolling in. Everybody was happy and particularly Matilda. To him the apparently limitless supply of this special species of all but naked, hairless kangaroos with boxing gloves on their paws for him to zap, made life a lark.

The bout was a crucial one. If Matilda got by Mickey Young there was only one more first-class middleweight, Cyclone Roberts, to stand between him and forcing Dockerty to lay his title on the line.

The contest had originally been scheduled for Philadelphia but the boxing salons had suddenly become stuffy.

The fact was that in no State of the Union where there was a law permitting professional boxing did the code specifically forbid a man fighting a kangaroo, for the simple reason that nobody had ever thought of it. When it came to staging a match with Matilda, Ahearn, Bimmie and Baker were at the mercy of spur-of-the-moment decisions by the Boxing Commissions which, legal or not, had the force of their authority. They simply said, 'No,' and to secure a reversal of this would have meant procuring a Court order or bringing a suit, with its endless delays.

Fortunately when the Pennsylvania Commission refused, the New Jersey fathers just across the river from Philadelphia, with an eye to both taxes and publicity, had proved more amenable and the bout had been set up to take place in the Downtown Arena, Camden, easily accessible by bridge from Philadelphia. This would result in the presence of the press from New York as well.

Patrick Aloysius came back to the ringside as the round ended. He was as gaudily clad as Bimmie but was looking grim and angry, his jowls shaking.

Bimmie asked, 'Something wrong? You look like you had bad news.'

Ahearn waited until Matilda and the Kid were at it again and under cover of the shouts of the spectators said, 'That was Sammy from our office in New York. They're bringing down Evil-Eye Finkel for the fight, to put the whammy on Matilda—give him the works. Sammy got it from one of the fellows who used to work in Dockerty's corner. Pinky's made the deal with Evil-Eye.'

Bimmie asked, 'Who's Evil-Eye Finkel?'

Patrick Aloysius looked at Bimmie with pity until he remembered that the boy was still comparatively new to the game. He explained, 'He's a son-of-a-bitch who hangs around boxing who claims he can put the evil eye on a fighter. He gets in the other corner and gives the boy the eye.'

Bimmie said, 'You're kiddin', ain't you?' but to his alarm Patrick Aloysius was looking doubtful.

He said, 'The bastard put it on one of my stable at St Nick's a couple of years ago and the kid quit cold on the deck in the third, from a push on the ulna bone. He wasn't hurt; he was just scared. Sammy just told me he ran into this handler at the automat and he said they were paying Evil-Eye two hundred and fifty for the double whammy with the swivel thrown in.'

Bimmie asked, 'Who's paying—'

Ahearn replied disgustedly, 'Who the hell do you think? The same outfit that has been on us right along.'

Ahearn's seriousness was beginning to alarm Bimmie who said, 'But what can happen— I don't get it?'

Patrick Aloysius elaborated:

Max (Evil-Eye) Finkel was a small-time ticket speculator with a hole-in-the-wall office who made enough out of his sideline to keep him in little luxuries as well as the public prints. His presence at the ringside was always fully reported.

His speciality consisted of putting the whammy, or hex, on a fighter by transfixing him with a stare from a pair of bulging eyeballs, one of which had the extraordinary capacity of being able to swivel while its mate remained static. He was open to engagement for an evening and for fifty dollars he would give them the left eye; for a hundred, the right eye; for two hundred both, and for two hundred and fifty, two plus the swivel, the latter blight being considered irresistible.

Ahearn finished his recital with, 'It's all a lot of crap.'

But by now Bimmie was even more worried. He asked, 'Did you ever try out using this schmo?'

Patrick Aloysius had the grace to look shamefaced. He said, 'Yeah.'

'What happened?'

'We win.'

Bimmie himself was too tough and shrewd a little New Yorker to believe that Finkel and his hypnotic and swivelling eyes had supernatural powers. All the hypnotist acts he had ever booked were phoneys. But if Pinky Schwab or Uncle Nono, or whoever was behind this move had taken the trouble to export Finkel to Camden, the menace was not to be ignored. He had no doubt that the advertised presence of this practitioner at the ringside could throw a fighter at whom the eye was aimed off his stride; most of them were superstitious anyway. African witch-doctors always contrived to let the victim know that a spell had been cast. Here a wizard was present in full view at the ringside. But would it work on a kangaroo? He queried Ahearn, 'Could Matilda go for this whammy?'

Patrick Aloysius replied, 'That's what I wouldn't know.'

'Can't we stop him—keep him away?'

Ahearn said, 'How can you? They buy him a ringside seat behind their corner, or sneak him into the Working Press. All the newspaper men write up that he's there. It's a great gag for them. When the bell rings he gets up and gives the other fellow the eye. See, the boy knows he's there and can't help looking just once. As I said, it's all a lot of crap, but...'

After the work-out, Bimmie took up the problem with Baker, 'What would happen if you were to look Matilda right straight in the eye, like? Like maybe you was going to hypnerize him, see?' And he gave a slight illustration by transfixing Baker's with his own blue pair, in a steady gaze.

Billy Baker shook his head, 'He wouldn't like it at all,' he replied. 'No animal does. Ain't you ever noticed that no animal can look you straight in the eye? He'd turn his 'ead away, he would. I tried it a couple of times like, and I saw he didn't like it. He won't look me in the eye and I'm like his brother. Why?'

'Oh, nothing,' Bimmie replied, 'I was just asking.'

When Bimmie had Ahearn alone later, he said, 'Did you hear what Baker said? It's just like those stories of a wild animal escaping from a circus and coming at you, and you look him right in the eye and he lays down and rolls over. We got to find an anecdote.'

'You mean an antidote,' Ahearn said.

'Yeah, an anecdote. So that when Finkel gives him the eye and Matilda looks, it won't work.'

Ahearn turned sardonic, 'Sure,' he said, 'you think of one. You heard what Billy said, what Matilda does when he looks him in the eye. If Matilda turns his head and drops his hands, Mickey Young got to belt him one, don't he? This is the one time where maybe Evil-Eye ain't a lot of crap.'

Bimmie said, 'Maybe we ought to call off the fight.'

'With Parkhurst coming down here and them New York writers and the fellows from Philadelphia?'

Bimmie shuddered. It was too late to cancel the fight and if suddenly Matilda were to succumb to the gaze of Evil-Eye, turn his head away and be flattened by his opponent in front of the collected press, it would be the end of everything.

Furthermore, as Patrick Aloysius had been at pains to

point out, Camden, an industrial town with much foreign-born labor, was something of a Mafia stronghold. The three just did not have the political muscle to keep Finkel away from the ringside. The warning had actually come too late.

The morning of the day when the fight was to take place Bimmie was still racking his brains for a means of foiling this attack, to find some way of neutralizing Evil-Eye's stare. And what had been driving him crazy ever since the problem had arisen was that there was something rattling around in the back of his head having to do with just that—neutralizing the whammy—and which he could not quite catch; a memory which he somehow connected with his schooldays, meagre as they had been, some ten years back.

Unable to sleep, Bimmie arose early and walked the streets of Camden. They had leaped too many hurdles, overcome too many obstacles on their way up to be ruined now by the simplest of all the plots directed against them.

He found himself passing in front of a toy store and stopped there for a moment, why he did not know, except that toy stores always had an attraction and he gazed at the collection of spaceman suits, dolls, tricycles, tennis rackets, air rifles, baseball bats, balls and gloves, dinkie toys, Indian and cowboy costumes—that idea was still trying to break through and then, pop! Almost like an explosion inside his skull, there it was. He had it! Ten years ago and today—the chances were a thousand to one he would be able to find what he wanted, but it was the only one and he had to take it.

He flagged a cab, drove over to Philadelphia and telephoned Paddy Aloysius from the station. He said, 'Looka, I got a idea. I'm at the station. I'm going up to New York. Never mind what it is, it's the only chance we got. If I find the guy I'm looking for, maybe I got the anecdote.'

'Listen, we ain't got time,' Patrick Aloysius said, 'the fight's tonight.'

'I know. I'll be back. Look after Matilda and cross your heart it should come out like I think. Don't say nothing to no one.' He hung up, caught the nine o'clock train for New York, arrived there at eleven, climbed into another taxicab and gave an address deep in the heart of New York's loft and commercial section, on the Lower West Side.

When the cab drew up in front of the Acme Paper Cup

Company, he got out, paid it off, went inside and trying to control his nerves and the trembling of his knees, asked the disinterested receptionist, 'Miss, can you tell me does B.B. Bucholz still work here?'

The receptionist decided the visitor was not worth rearranging her hair for and merely asked languidly, 'B.B. who?'

Bimmie started to repeat and then stopped and corrected himself, since it was unlikely he would be known there by his P.S. 187 nickname. 'Lester Bucholz. He's a big, tall skinny guy.'

The Paper Cup Company was a large firm. The receptionist looked through a list of names and numbers on her desk and to Bimmie's enormous relief said, 'Yeah, I think he works in packing,' and jerked her head in the direction of the corridor which opened off from the office. Bimmie wandered down it until he came to a glass door labelled 'PACKING DEPARTMENT'. He went in. A dozen or so hopeless young men were listlessly pushing stacks of paper cups into cartons designed for vending machines. It was not an operation that took a great deal of brains or held out much of a future. One of them at the far end, tall, gangling and now sporting a mod, drooping moustache and long sideburns was indeed Bucholz.

Bimmie said, 'Hello, B.B.!'

Bucholz looked up without enthusiasm and said, 'Hello, Bimmie!' his fingers automatically continuing to stuff paper cups into a long, cardboard container.

'How's things?' Bimmie asked.

'Okay.' It was the voice of a thoroughly defeated man. Then he added, 'I seen you're doing all right. Had your pitcher in the papers, with some kind of an animal like a seal. What is it? You in the circus?'

'Nah,' said Bimmie, beginning to wonder whether his mission was going to be a success. 'It's a kanagaroo from Austeralia. He's gonna be a champeen. I was just lucky.'

Bucholz said, 'Yeah, I guess so. You want something?' His fingers completed another carton and stacked it up on the side with its brothers by the hundreds. Other fingers all around were doing the same.

Bimmie said, 'Yeah. Maybe I would.'

Bucholz did not look interested. He asked, 'How did you know where to find me?'

Bimmie said, 'I heard you got a job here when we quit school.' He did not add that the estimate he had made of B.B.'s intelligence at that time was that whatever job he was able to get and whatever it was he started at, it was even money he might still be doing the same thing. He was pleased that his judgement had been correct, and that he had not had to waste time tracing Bucholz from place to place.

Bimmie said, 'Listen, B.B., do you remember the days at P.S. 187?'

'Yeah,' said Bucholz, 'they stank. What's that got to do with anything'

Bimmie asked, 'Do you remember what you used to do that used to drive Mr Abrams the math teacher and fat Miss Purdy from the English Department, nuts, and the kids too?'

For the first time a gleam, a faint spark, showed in the sad, dull eyes of Bucholz. 'Yeah, yeah,' he said, 'that's right! Yeah, sure! Boy, did that Miss Purdy jump.'

Bimmie asked the sixty-four-dollar question, 'Can you still do it, B.B.?'

The lack-lustre eyes finally came wholly to life, 'Yeah, yeah, sure!' he said.

'Can you show me?' Bimmie asked.

Bucholz was now fully awake, his fingers had even ceased from automatically functioning and he said, 'Can I! Funny thing, now, you remembering like, after all these years. I drive 'em nuts here, too.' He turned his back for a moment.

At the end of the long individual table where Bucholz was working, filling and stacking cartons, there was a half-empty tumbler of water from which someone had obviously had a drink, the owner of the business not being one to encourage the use and waste of his own paper cups.

'You want an idea?' The glass was about ten feet removed from where they were standing. Bucholz turned and smiled at it. With a shivering tinkle the glass disintegrated into a dozen or more shards.

One of B.B.'s fellow workers said wearily, 'Oh for Chrissake, Bucholz, cut it out!'

'Boy!' said Bimmie.

166

'Better than ever,' Bucholz boasted.

Bimmie went over and inspected the pieces of broken glass, shaking his head in admiration. 'Boy, oh boy!' he said. The sinking feeling had departed from the pit of his stomach; his heart was singing within him. He said, 'Listen, B.B., can you get off for the rest of the day?'

Bucholz shrugged, 'I dunno. What for? I got a job here.'

'There's a "C"—a hundred—in it for you.'

Bucholz brightened again, 'Yeah? What's it all about?' The rest of the workers at the mention of this munificent offer had stopped poking cups into cartons and were listening.

'Where can we talk,' Bimmie asked.

'In the can,' Bucholz replied. He led the way to the employees' lavatory and they went and locked themselves in.

Bucholz said, 'Okay, now what's it all about?'

Bimmie told him, omitting however to refer to the nature of the parties responsible for engaging Evil-Eye. He did not know how steady B.B.'s nerves might be and did not want him handicapped.

Bucholz reflected and then asked, 'Where would I be? How far? I never been to a prize fight before.'

Bimmie replied, 'I'll have you right up at the ringside in the Working Press. Maybe no further away than like that glass was. They're paying the son of-a-bitch two-fifty to give my kanagaroo the left, the right and the swivel.'

Bucholz's large ears seemed to go up like those of a bird dog and Bimmie realized he had made a mistake in mentioning Evil-Eye's fee, an apprehension that was confirmed immediately when Bucholz replied, 'Maybe I should get more.'

Bimmie, however, was in no mood to haggle. He said, 'Listen, B.B., you do your job and I'll pay you the same like they're paying Evil-Eye.'

Bucholz nodded and then asked, 'I wouldn't get into no trouble, would I?'

'Nah,' said Bimmie, 'not if you keep your mouth shut afterwards.'

Bucholz said, 'Okay, maybe I ought to get in a little practice,' and he turned towards the mirror over the wash

167

basin, but before he could smile at it, he was stopped by Bimmie's yell.

'Jesus Christ, no! You want we should start off with seven years of bad luck? Ain't I got enough trouble? You can practise when we get down there, but not on mirrors. Come on, maybe we could catch the two o'clock.'

He had a moment's thrust at his heart. Here he was in New York and Hannah at the other end of a telephone. He wished he might ring her or even see her. But that was all over, he remembered. She had given him the air because he was a little crook, but what kind of honesty was it to send down a witch-doctor to put the hex on an innocent animal? What kind of a world was Hannah expecting to live in? He guessed there was no use calling her.

They came out and B.B. said to one of his associates, 'Tell the boss I got took sick sudden like.'

'Honesty!' Bimmie said to himself, with some bitterness. 'You can't get away from it.'


The New York press was there, in addition to Parkhurst and O'Farrell, Junius Jones, Petrie, Hobart and Horne were out in full force, their editors finally having heeded the old adage, 'If you can't lick 'em, join 'em and if possible take the story away.' The presence of Evil-Eye Finkel added a fillip and there was the further rumor that if Matilda got by this and his next match, something really big might be brewing.

Duke Parkhurst had jammed his huge bulk into the narrow Working Press seat in happy anticipation. Birdie, who he had brought down with him, was placed a few rows behind in a ringside seat. It was the first prize fight she had ever attended and she was making a valiant though not wholly successful effort to understand what was going on. To the six thousand patrons who packed the floor and the mezzanine of the tidy, newly-built, indoor amphitheatre just the excitement of being there was already worth the money, whatever might happen.

The preliminaries were done and the main event was in the ring. Billy Baker was putting the gloves on Matilda while Bimmie distracted him with his usual pre-battle banana.

Junius Jones leaned over to Parkhurst and said, 'Duke,

you crazy bastard, I would never believe it if I weren't seeing it. Can that thing really fight?'

Parkhurst said, 'You wait!'

The crowd was humming and seething with nervous tension. At that moment a large, stout man arose in the second row of the Working Press benches. He had a moon face, thick, baby-pout lips and black hair cut shoe-brush style. He was wearing dark glasses. As he stood up, he removed them to reveal the extraordinarily distinguishing part of his features, a pair of bulbously protruding eyes, one of which appeared to have a cast in it. He gazed around the house and to the horror of some of the ladies in down-front ringside seats, one of his eyes—the one with the cast—swivelled ninety degrees, while the other remained quiescent.

'Holy cats!' cried Gill Hobart. 'It's Evil-Eye. He's really here.'

'Looks good,' put in Jack Horne.

Petrie called over, 'Hey, Finkel! Which one are you giving him tonight!'

A local sportswriter enquired, 'Who the hell is Finkel?'

The New York contingent enlightened their Philadelphia colleagues. A group went over to interview him.

No one took any notice of a tall, thin, melancholy fellow whose moustache and long sideburns gave him somewhat the air of the sorrowful Knight of La Mancha, seated in the Working Press a dozen or so feet away from boxing's only accredited witch-doctor. He, too, rose but only for a moment. Then he resumed his seat and busied himself with his program.

The reporters swarmed around Evil-Eye.

'Are you using the swivel tonight, Evil-Eye?'

'Is it true that you can paralyse somebody with a look?'

'Tell us about it, Evil-Eye.'

'Who was the last guy you put the whammy on?'

This, of course, was up Evil-Eye's alley, his method of advertising his presence and a part of whatever power he possessed to irritate, distract or intimidate his victim, and he was heard to reply, 'Tonight I'm going to give all three eyes; the left one, the right one and the two together with the swivel on top.'

169

Another reporter asked, 'But will it work on a kangaroo? He ain't never heard of you.'

Evil-Eye held up one hand and said, 'Don't come too near to the eyes, boys. I'm working up the power and I wouldn't want to hurt any of my friends.' To guard against any such catastrophe he lowered the dark glasses again and said, 'I ain't never failed yet. An animal is of a lower classification than man. He is not possessed of the same brainpower or resistance. With an animal it ought to be duck pie. In my opinion, three eyes is a waste. I could do it with one. Three he's getting. It shouldn't last more than a couple of minutes.'

Duke Parkhurst had a sudden qualm. Not that for a moment did he believe that Finkel's eyes generated any power greater than being able to recognize a bagel at three paces. But it did strike him that if Matilda looked down and found himself confronted with this revolting countenance with its grisly movable orb, he might be disconcerted just sufficiently to have his mind taken off his work, in which case Duke was very well aware who would be the laughingstock back home. For a moment he contemplated arising, seizing Evil-Eye by the folds of his fat neck and ejecting him bodily by force, but then he realized that he would be making a fool of himself. Besides which, it was too late. The introductions, the instructions from the referee had been completed; the two contestants were in their corners awaiting the bell. Matilda as usual when erect, suddenly appeared extraordinarily menacing and, as always, the crowd gave vent to an awed shiver and murmur of anticipation.

The bell rang and simultaneously, as Matilda took his first hop to the center of the ring, his left out, his right covering the side of his long muzzle, Evil-Eye Finkel stood up and removed his dark glasses.

At that same moment the lanky, gloomy-looking stranger in the Working Press, sitting a few yards away, got up, faced Finkel and smiled.

The reports of what followed were so varied and confused that no two observers of the affair were able to furnish the same description or, for that matter, take in the entire scene as Evil-Eye Finkel clapped one hand to his face and screamed, 'Ow! My eye! My eye!' The tall stranger smiled again and immediately afterwards Finkel's other hand went up and the screams were augmented.

'Oi!' bellowed Finkel, 'My eyes! My eyes! Help! I've been shot! Get me a doctor!'

Sportswriters sitting close by went to the assistance of the stricken man who was unquestionably in agony. They were unable to pry his hands away from his face or stem his piteous cries of 'My eyes! My eyes! I'm blind! Get me to hospital!' to the point where it seemed that an incipient panic might be in the making at the ringside.

Special police rushed around to offer help and search out any gunman who might have been responsible for the unauthorized use of firearms, but then nobody had heard a report or an explosion of any kind. Spectators were beginning to stand up and crane their necks at the commotion. The bulk of Duke Parkhurst loomed over all as he arose, unable to figure out what had happened, but filled with a curious sense of, if not divine, then some other kind of justice that chose to select as its target Finkel's own weapon—his eyes. Whatever their power or their effect upon Matilda might have been, no one was going to discover that night for they were definitely out of commission.

By now Finkel, moaning and sobbing, had collapsed backwards over the Working Press benches and the doctor always in attendance at a boxing bout hurried to his side. A woman screamed and four of the special cops provided by the management to keep order, quickly picked up the writhing figure of Finkel and hustled him from the arena into one of the dressing-rooms for further examination and treatment.

Such was the excitement, turmoil and hullabaloo about the ringside caused by whatever it was that had happened to Evil-Eye, that hardly more than a handful heard the sound of a crack like the breaking of the branch of a tree, followed by what even the most literate of the reporters was compelled to describe as a dull and sickly thud. And only then were glances turned to the combat area within the ropes, where it was seen that Matilda was standing in a neutral corner while the referee was completing the count over the inert figure of Mickey Young who, according to one of the few witnesses

171

who was able to report the punch, had after one minute and eighteen seconds run afoul of one of Matilda's short, right-hand chops and had fallen face forward like one struck by lightning.

Such was the drama outside the battle zone that nobody really had time to be impressed with Matilda's performance, particularly when the doctor returned to the scene announcing, 'Somebody shot that man in the eyes!' Cupped in the palm of his hand he exhibited several number .006 size buck shot.

More excitement as the chief of the special police said to his minions surrounding the Working Press section of the front row, ringside seats, 'Don't nobody leave here until we go through you.'

Reporters and spectators were compelled to stand up and one by one raise their arms while the police did a quick slap-slap frisk of their side and rear pockets and possible shoulder holsters, but no lethal weapon of any kind was found.

Parkhurst asked of the doctor, 'Is he badly hurt?'

'No,' replied the medic, 'but he could have been. The bullets were apparently spent. I don't know what those flatfeet are doing searching fellows around the ringside. The shots must have come from the balcony somewhere, maybe with one of those rubber band catapults. He's only temporarily blinded from the bruise. He'll be all right in a week or so.'

Parkhurst looked up at Bimmie who was leaning over the ropes and said, 'What happened?'

Bimmie said, 'A left to the belly and a right to the chops. Mickey Young turned his head a minute to look what was going on. A fighter oughtn't never to do that.'

'No, I mean to Finkel.'

'I wouldn't know,' Bimmie replied and then added with a sanctimony that did not quite become him, 'Maybe it was the hand of God that struck him down.'

The frisk finished and fruitless, the crowd was dispersing, buzzing happily. They had not seen much of a fight or much of what fight there had been, but they had been present at a thoroughly exciting evening in which, as they embellished it, a man had been shot to death before their very eyes at the ringside.

Duke Parkhurst's telegrapher was looking at the big columnist in anticipation of seeing magic words decorate the sheet of yellow copy paper in his typewriter, but none were forthcoming. The Duke was sitting there thinking. Somewhere, something had managed to elude him. It had all happened just too pat: the moment of retribution, its nature, the size and calibre of the missiles, the doctor's report. He said to his operator, 'Hang on a minute. Keep the line open. Message the office, 'I've gone to check something.' He heaved himself out of the narrow press benches and motioned to Birdie who had a handkerchief pressed to her lips.

She said anxiously, 'Oh, how awful! Did somebody shoot that poor Mr Young?'

'Practically,' Parkhurst said. 'Come on, we're going to pay a little visit and see where that smell of defunct rodent comes from.'

They made their way to the dressing-room of Bimmie, Matilda and Company where they found them in a gleefully back-slapping celebration. This centered not on Matilda, who was hunkered down in a corner contentedly munching his Giant Jumbo Almond-Studded Hershey Bar, but instead around a tall, moustached, sideburned stranger who looked like—Parkhurst's trained photographic memory flashed him—someone he had seen at the ringside before, during or after the fracas, or all three.

The celebration stopped abruptly as the columnist entered and Bimmie cried, 'Come in, come in, Mr Packhurst! Say, wasn't Matilda great? Did you see that punch?'

Parkhurst said, 'Never mind Matilda, who's this guy?'

It was time for a flow of schmoos, but somehow in the presence of Parkhurst and his manner, it would not come. Bimmie just said, 'Just an old friend of mine.'

'What's his name?'

'B.B.—I mean, Lester Bucholz. Lester, shake hands with Mr Packhurst, the greatest sportswriter in the world.'

But it was too late. Parkhurst had already got the

nickname and his mind computerized to every form of fight-game shenanigans had already performed the operation.

'Give!' he said to Bimmie. 'What did he hit him with and how did he do it?'

For a moment more Bimmie was able to maintain his bland air of innocence as he said, 'I wouldn't know what you're talking about, Mr Pockhurst...' until the newspaper man pointed a finger as thick as a gas pipe at him and said, 'Listen, you little bastard, I said give! If it weren't for me, you and that kangaroo wouldn't be here. I made you and I can break you.'

Bimmie's native shrewdness asserted itself. He knew when he was licked. 'See,' he explained, 'B.B. and I were at P.S. 187 together when we were kids, and B.B. had this trick, he could put a BB shot between his teeth—that's how he got nicknamed B.B.—and give a kind of a click and shoot it out like a bullet. He used to hit Miss Purdy, our English teacher, in the ass—excuse me, miss—with it when she turned to the blackboard and I'll bet she don't know to this day where it come from. B.B. was a dead shot at ten yards with them pellets and could take a kid in the back of the neck half-way across the room.' He turned to Bucholz, 'B.B., show Mr Pockhurst.'

Bucholz asked, 'Is it okay?'

'Sure, sure,' Bimmie promised him, 'he's a friend of mine.'

A calendar hung suspended on the opposite side of the room, one which had been got out and donated by O'TOOLE'S BREWERY, CAMDEN, N.J., which legend was printed across the top over the thirty-one days of the month of October. Bucholz appeared to be searching with his tongue in his cheek for something and then he smiled at the calendar, or rather he exhibited a row of strong, horselike white teeth. There was a faint popping sound and a small round hole appeared in the middle of the first 'O'. Again a grin and a second hole appeared in the second 'O'. A third pop and the final 'O' was drilled dead center.

'See?' Bimmie was going on. 'So when they said Evil-Eye was coming down here to put the whammy on my kanagaroo, I remembered B.B. from like when we was at school together and how he could knock a piece of chalk out of your fingers at the blackboard, and went to New York to look him up, and got him to come down and . . .'

But Duke Parkhurst was no longer listening to him. He went over to the calendar and inspected the three neat holes carefully. Then he leaned down and from the floor where they had fallen, retrieved three small BB shot in his palm, while a most beatific smile spread over his large countenance. He went over to Bimmie, shook his hand and did the same to B.B. Bucholz, motioned to Birdie and then without another word, left the room.

Patrick Aloysius, who had taken no part in this conversation but had been quietly sipping his glass of Jameson's Irish, suddenly frowned and asked, 'Who was the chick with the Duke?'

Bimmie said, 'I dunno, but wasn't she a sweetheart? Boy, did you see them eyes! He can pick 'em, can't he?'

Ahearn took a good pull at his drink. 'I hope so,' he said, 'but I've seen that babe before somewhere.'

At the ringside Parkhurst was beginning to dictate his story. Several times the tough, wiry telegraph operator who thought he had seen and heard everything, stopped and stared at him and said, 'You ain't kiddin', are you?'

The Duke repeated his litany, 'Parkhurst, he don't kid. I'm telling it like it was.'

# XVII

IN ADDITION TO the destruction of the mystique of Evil-Eye Finkel and his whammy forever—the Broadway witchdoctor was never seen at any fight club again—the Camden affair had a number of repercussions of varying importance worth noting in the saga of Matilda.

One of the happier consequences was the elevation of B.B. Bucholz from an unhappy nonentity into something of a celebrity.

The publicity he achieved via Parkhurst's story of his unique skill led to a four-weeks' engagement in a night club giving exhibitions of his incredible marksmanship with the tiny shot impelled by his teeth, in the finale of which he played 'My Country 'Tis of Thee' by striking variously tuned tumblers with the pellets. Naturally Bimmie booked the act and in a rush of gratitude cut him only ten percent. Later, in further acknowledgement of his public sensation, the Acme Paper Cup Company promoted him from the packing to the shipping department with a ten-dollar raise, so that all in all B.B. Bucholz enjoyed a better life than he had before.

Less satisfactory was the outcome when, on the wings of euphoria engendered by his triumph, Bimmie attempted to contact and make it up with Hannah.

Back in New York, he did ring her on the phone the day after the appearance of Parkhurst's account of the happenings in New Jersey and thrilled to her voice at the other end.

'Hannah, baby!' he said. 'This is Bimmie. I'm missing you terrible. I'm in New York. I guess you read all about me.

Can't we make it up? I didn't mean no disrespect about your old man—I mean, Mr Lebensraum. A lot of mark-ups I know are worse for a lot less and I didn't take into account his overheads and expenses, which ain't for nothing these days when you run a class store like his. You shouldn't hold it against me, like, maybe when I'm talking my head off because I'm so crazy about you. I'm going places, Hannah, and maybe something big is coming off. So big I can't even talk to you about it on the telephone. But if maybe you'd let me come up to see you and I could explain...'

He stopped there, not because he had run down by any means, but to see what the reply would be, if any, and found himself saddled with one of those impregnable silences at the other end of the phone about which one can do nothing since there is no way of either turning on the charm or taking someone's hand. He said, 'Hannah, baby, are you there?' and at last her voice filtered back.

'It's no use, Bimmie, I told you how I felt about being honest. I think it was awful what you did to that poor foolish man. You might have blinded him for life. Please don't call me any more, Bimmie,' and thereafter the connection was broken, leaving him desolate and depressed, a melancholy not lifted even by the memory of the conference at which he had been present in the office of Parkhurst, embracing Justus Clay the managing editor, Parkhurst and himself where a project was discussed to which not even Patrick Aloysius or Billy Baker were to be made privy until further developments.

For the Camden fight had broken the log jam. Matilda no longer existed only in the column or on the sports pages of Duke Parkhurst in the *Mercury*. He had now been acknowledged by the other newspapers as well. True, some of the stories were written humorously or ironically with *l'affaire* Finkel rather taking precedence over the actual bout, but what emerged and particularly in the more sober account of Cassius Jones of *The Times,* was that Matilda, a trained kangaroo had disposed of the third-ranking middleweight of the United States with two perfectly delivered punches in the opening stanza of what had been billed as a ten-round contest.

Parkhurst and Clay reviewed these stories in Clay's office, including those of the Philadelphia and Camden press, with the utmost satisfaction. In a reversal of the usual newspaper policy this was no longer a story they wished to keep exclusively for themselves. They wanted it spread as widely as possible. With his inside track to Matilda and Company, Parkhurst would always have an edge on the others. What was important was that they had succeeded in establishing Matilda as a serious contender.

Mr Solomon Bimstein was announced and admitted to the sanctum. He had by now discarded the mustard-colored hat and the outfit that went with it, as with increasing prosperity his tastes had also become somewhat more educated, and he could afford to copy some of the better dressed denizens of Mazda Lane. Vicissitude coupled with success had also matured him somewhat. He had the sense to realize that he had been admitted to the very base of the footstool of God, and thus took the seat indicated by the jerk of Parkhurst's head and remained there quietly waiting to speak until he was spoken to.

He was not too much encouraged by the sour expression on the simian features of the managing editor, or the up and down he was getting from Parkhurst. Finally the latter spoke.

He said, 'Bimmie, we've sent for you without your partners because there's something we want to discuss with you and we think you can keep your mouth shut. You don't drink so you're not likely to spill the beans in some bar. Also you've been in show business long enough to know that premature exposure of a big deal can wreck it. When that happens everybody wants to get into the act and this is one we want to keep exclusively to ourselves. Mr Clay and I would prefer that you don't discuss this with either Patrick Aloysius or Billy Baker and certainly not with anyone else.'

Bimmie, looking from one to the other, 'Numb's the word.'

'Another reason we're letting you in on this,' Parkhurst continued, 'is that you showed imagination in your handling of Finkel and Company.'

Bimmie accepted the compliment in silence. Schmoos was

for when you were in trouble. Obviously at that moment he was not.

'You may be called upon for more of the same—if not imagination, then ingenuity. Your next bout is scheduled with Cyclone Roberts.'

Bimmie said, 'Matilda will take him like Grant took Atlanta.'

'Richmond,' murmured Clay half to himself, editing being vocational with him to the point of obsession.

'I don't doubt it,' Parkhurst said, 'and that isn't what I'm worried about. It's getting the fight on in Chicago. Have you tried yet?'

'No sir, but after the way we got permission in New Jersey there oughtn't to be any trouble. Paddy Aloysius and I are going out there next week to talk turkey.'

'Listen to me, Bimmie. It isn't going to be that easy, at all. The Chairman of the Boxing Commission in Chicago is an ex-amateur heavyweight college boxing champion and a gent. He's also got a very long, hard nose. I don't think you're going to get by him.'

'Well, if we don't,' Bimmie said, 'we can bring the fight to New York, where...'

'No you cannot!' Parkhurst said. 'We're saving up New York for something else. Mr Clay and I have brought you in here to tell you that you've got to stage that fight in Chicago.' And then told him why.

When Bimmie left the office his knees were shaking so that he could hardly walk. He took at least four wrong turnings before he ever managed to reach the banks of elevators to take him to the street, where he also came that close to walking under (a) a bus, (b) a taxicab, and (c) a beer truck. He remembered once Parkhurst's admonition against thinking in terms of peanuts but even he found it difficult to visualize in concrete terms the riches that Parkhurst and Clay and the *Mercury* had held out to him, a million dollars or more.

On the other side of town, high up in his eyrie Gio di Angeletti sat in his elegant office at his elegant, ebony desk, his feet sinking into the elegantly thick carpet pile and with

179

his usual mixture of emotions was rereading Duke Parkhurst's account of the débâcle in Camden. He was irritated and at the same time he wanted to laugh but Johnny Renato his right-hand man was with him. It was the fact that Renato was so totally lacking in humor, this void being filled by unswerving loyalty and efficiency, that made him such a valuable henchman. But di Angeletti had to keep his face straight in his presence.

The *Capo* asked, 'Who was actually responsible for this?'

Renato said, 'Well, Pinky dug up this Evil-Eye fellow; he said it had worked for him once. But Joe Marcanti okayed it. You know how they feel about the evil eye in Sicily—they put a lot of faith in it. I suppose that makes it his responsibility.' Then he added reflectively, 'Marcanti's been screwing up this deal ever since it started. Maybe if he was hit just a little, say he fell somewhere and broke his leg...'

Di Angeletti shook his head and said, 'No, not yet. It wasn't his fault. Any Sicilian who doesn't believe in the evil eye isn't worth his salt. Besides which it might have worked, if they hadn't found this—What's his name?'

He consulted the page of the *Mercury* featuring a shot of B.B. Bucholz in action and choked. But then as he reread Parkhurst's story and caught its sense of malicious triumph between the lines—and not always between—the desire to laugh faded. This was the real enemy who was threatening to make a fool of him. The small fry were not worth bothering with for the moment. He wondered whether Parkhurst had any inkling that the hunter was being hunted. He asked Renato, 'What about Birdie? Was she there?'

'Yes.'

'They stayed overnight?'

'Yes.'

'Where?'

'The Quaker State Hotel in Philadelphia.'

'Any pictures?'

Renato produced two photographs from an envelope, two excellent sneak, candid shots that showed Parkhurst and Birdie at the registration desk of Philadelphia's premier luxury hotel.

180

'That's nice,' Uncle Nono said. 'Birdie looks good. What was on the book?'

'Just "Parkhurst". He reserved the room by phone that afternoon.'

'The house dick?'

'Who would bother Duke Parkhurst?'

Gio di Angeletti contemplated the photographs again and said, 'Lovely! Any affidavits?'

Renato said, 'The maid and the floor waiter,' and produced two papers. Then he asked, 'Shall we. . .?'

Uncle Nono reflected. One night in a Philadelphia hotel? It wasn't good enough, nor was it the time or the place. When you had a man by the short hairs you waited for just the right moment to bring him down. 'No,' he said, 'not yet. He's gone once, he'll go again. If the F.B.I. had to pull every guy who took a dame on a little trip once in a while, they wouldn't be doing much else. It's when somebody begins to make a habit of it that they get interested. They really wrote the law to break up white slave rings and occasionally they like to remind everybody that it's there.'

At the end of Parkhurst's story there was a brief paragraph to the effect that Patrick Aloysius Ahearn had signed a contract with Whitey White, the manager of Cyclone Roberts, for a match that it was hoped would be staged in the Chicago Coliseum. Di Angeletti frowned and asked, 'What about this Chicago match?'

Renato said, 'Forget it, boss. They'll never get by the Illinois State Athletic Commission.'

'They could take it somewhere else.'

'Not after they'd been turned down in Chicago. The Commission there pulls a lot of weight. If you ask me, it's the end of the line.'

'What if Walter Mason wanted to stage it here in the Arena? It would be a sell-out! And after all this publicity he knows it; he isn't in business for his health.'

Renato said, 'I wouldn't worry about that. With Wild Bill Wildman in your pocket he wouldn't dare. He knows he's in the doghouse already for giving that fellow Bimstein a manager's licence.'

181

Di Angeletti nodded, 'All right,' he said, 'but it wouldn't do any harm to let Marcanti know that his name came up.'

After Renato had departed, Gio di Angeletti leaned back at his desk and reflected upon the retributive disaster that had befallen Evil-Eye Finkel and permitted himself the chuckles he had been compelled to suppress.

The three members of the Illinois State Athletic Commission seated around three sides of a table, were regarding their visitors with extreme distaste. The latter consisted of Bimmie, Patrick Aloysius and Billy Baker.

Two members of the Commission were old-time, veteran ward politicians, the third, Payne Waddell, the chairman, was a younger man, a graduate of the University of Illinois where he had been intercollegiate heavyweight boxing champion, a lawyer and a gentleman. He sat as though he could not believe the words he was hearing. 'You mean to say that you propose to stage a ten-round bout between a former middleweight champion Cyclone Roberts of Peoria and a kangaroo?'

Bimmie said, 'Yessir, you see...'

The chairman interrupted, 'And you're actually asking us to issue you a licence to produce such an affair?'

'Yessir,' said Bimmie, 'because you see...'

But he was cut off again by the chairman's withering glance and query, 'Just what do you take us for, Mr Bimstein?'

This time Bimmie was able to launch and gain a few yards, 'You see, sir, this ain't a ordinary kind of kanagaroo. He knocked out Lee Dockerty, middleweight champion of the world. He ain't like the usual boxing kanagaroo that you see in acks. He's got class. We already knocked out eleven more guys—the kanagaroo, I mean—and if we can get by Cyclone, Pinky Schwab got to give us a return match in New York, like he said in the papers, so all we want is a crack at Cyclone whose manager said he would fight us if we could get the okay from you gentlemen, and...'

'Well,' said the chairman, 'you can't.'

He looked at one of his fellow commissioners who said, 'I

don't know what these fellows have been getting away with. They've been fighting down south and out west and this kangaroo seems to keep knocking them over.'

'Well, he won't in Chicago,' said the chairman. And then to Bimmie, 'What's this, a joke or something, taking up our time? This is a civilized city. There's a circus playing in town, take your animal over there and maybe you can get him on a clown act.'

Bimmie reached inside the pocket of his jacket and produced already well-worn copies of Parkhurst's initial pieces. He said, 'But look here, sir, here's what Duke Pockhurst wrote about Matilda—that's our kanagaroo—knocking out Lee Dockerty in Mississippi. So Mr Pockhurst says that Matilda got to be the champion until Dockerty fights him again and can lick him.'

The chairman said, 'Duke Parkhurst isn't writing a column in Chicago, and if he was, it wouldn't make any difference.'

Paddy Aloysius had a go. He said, 'Bimstein, Baker and myself are all licensed in New York to act as managers or seconds...'

'Well, you haven't got a licence in the State of Illinois,' said the chairman, 'and what's more, you won't get one.'

'But Mr Chairman!' began Bimmie.

Waddell pointed a long, menacing finger. He said, 'Paddy Ahearn, I know you. You usually manage to keep out of trouble and you look after your boys pretty well. I'm surprised that you should be mixed up in a thing like this. You, Mr Bimstein and you, Mr Baker, I don't know and what's more, I don't want to. And the sooner you shake the dust of Chicago from your feet, the better we'll like it, and take your circus act with you. And don't try to sneak into some hole-in-the-wall club and pull this off, or we'll throw the book at you and you'll find yourselves explaining to the judge. The laws with regard to illegal prize fights in the State of Illinois are specific. We'll be keeping an eye on you and so will the police. We don't want that kind of a stag show in this city.' He looked at the two other commissioners and said, 'Isn't that about the way it is, boys?'

183

He was rewarded with, 'Sure, sure, Mr Chairman, that's right. Just like you say. We ought to kick their asses right out of the office and down the stairs.'

The chairman said, 'No need to go to such lengths. I think we all understand one another. You try to put on a fight like that in this city and I'll see that you spend the next six months in the cooler. Now, beat it! The interview is over.'

A few minutes later they found themselves on the sidewalks of Michigan Avenue.

Paddy Aloysius said, 'That Parkhurst knew something, didn't he? That's what happens when you appoint a gent to a job like that. You can't talk to him. We're cooked.'

Bimmie had a momentary and quite harrowing cartoonist's vision of a money sack on the side of which was written, '$1,000,000.' The sack had wings and was flying away. He was also remembering his last conversation with Parkhurst, and the call for him to use not only imagination but ingenuity. But what imagination and what ingenuity? It was all very well for Mr Pockhurst to tell him that the fight had to be staged in Chicago, but then Mr Pockhurst had not seen this chairman who even spat on Mr Pockhurst's column. Ingenuity-schmooety, what could a little ex-Broadway booking agent do against a big-shot Commission? This was not a case of digging up another B.B. Bucholz. Who did he know in Chicago? Nobody? Only maybe another booking agent for whom he had once done a favor, a Mr Herman Kaplan who had a couple of pop singers he got on television occasionally.

Patrick Aloysius was shaking his head gloomily. He said, 'The stuck-up-son-of-a-bitch! We just knock out Mickey Young and he says take your kangaroo down to the Circus!'

Something went click inside Bimmie's skull, just as it had the morning when he had stopped in front of the toy store and the air rifle on display with its box of BB shot had reminded him of Bucholz. Patrick Aloysius's remark had managed to impinge upon Bimmie's recollection of Herman Kaplan who booked clients for . . .

'What if we did that?' Bimmie asked of Ahearn suddenly, trying to slow down the pinwheels going around inside his skull. 'What if we did that, just like he said, and put Matilda back in a show? Could he do anything to us?'

'Certainly not,' Patrick Aloysius replied, 'a show is a show and a fight is a fight. They haven't any jurisdiction over shows. What the hell do you think you're going to do—put on Matilda and the Cyclone in the Center ring?'

'Oh boy!' said Bimmie. 'Have I got a idea! Mr Smartie Chairman who spits on Mr Pockhurst, you'll be finding out! Let's get back to the hotel where I can call up a friend of mine.'

Ahearn looked at Bimmie in disbelief. 'You mean you know somebody in this town who could cut Waddell down to size? What's his name?'

'Mr Herman Kaplan,' Bimmie announced proudly. 'He books acks for TV.'

'Oh, for Chrissake!' said Patrick Aloysius.

# XVIII

'SEE,' BIMMIE WAS saying, 'you make it a kind of a story, like. In the beginning where this kanagaroo is lost in the desert without any water or anything to eat, and can't find its mother. See, you play a kind of sad music along at that part. And this man comes along who has some sangwidges and water, which he gives to the kanagaroo and takes him back to where his mother has got lost and so he's very grateful. This part of the ack is short and don't take very long to show, but it's so you'll know later when it comes to the fight part how the kanagaroo is grateful to the man. Well, so, years later, this man who is a manager, like Patrick Aloysius here, is down and out and going broke. But he can get back in the game again if he can find someone who would fight this here now Cyclone Roberts, who nobody wants to fight because he's too tough. So then there's this circus which is in town, with a boxing kanagaroo ack and the manager and the kanagaroo recognize each other from the time when he saved his life in the desert.

'"Will you fight this Cyclone Roberts for me?" the manager asks.

'"Yes I will. For once you saved my life when I was lost in the desert and brought me back to my mother..." someone says into the microphone, for like now, our Matilda can't talk.'

'Never mind the dialogue,' said Mr Bursten, a sallow-looking man in shirtsleeves with a pencil stuck behind his ear. He was Production Manager of C.T.S.—Chicago Television Syndicate, the Midwestern network that blanketed seven of the most populous States of the Mid West.

'Well, see,' said Bimmie, 'you got to make it like a real story, like maybe Matilda is really talking.'

'Sure, sure,' Mr Bursten said, 'so then the kanaga—I mean, the kangaroo fights with this Cyclone Roberts, lifts the mortgage on the old homestead and marries the poor widow's daughter.'

'Wait a minute, Ivan,' said Fred Akely, Vice-President in charge of programming, an alert, sharp-eyed man who looked younger than he was. 'The idea is that somewhere along the line the viewer might get to see one hell of a fight.'

Ivan Bursten appeared shocked and asked, 'Are you kidding, Fred? What do we call it? Androcles and the Kangaroo?'

Another voice was now added, that of a stocky, dark-visaged individual, one of those whose chin was always blue in spite of a fresh shave. He was Herman Kaplan, Bimmie's friend who, it turned out, had become one of Chicago's most dynamic and prestigious booking agents. Smoke the color of his chin curled up from a Churchill-sized cigar as he said, 'Nobody's kidding. The way my friend here, Bimmie, has got it you can't miss.'

Bursten said sourly, 'The way your friend Bimmie tells it, we'd be out of business the next morning.'

Akely now turned to Patrick Aloysius and said, 'You say the Boxing Commission turned you down...'

'Flat!' said Ahearn.

'And threatened you with arrest if you tried to pull off the fight anywhere in Chicago?'

'Yep.'

'And you want us to pull your chestnuts out of the fire?'

Kaplan removed his cigar long enough to say, 'Chestnuts, my foot! We're just giving you first crack at a fight everybody wants to see. Cyclone went fifteen rounds with Lee Dockerty and had him on the deck once, so it's a natural. We're just showing you how it could be done. As a friend, I come to you fellows first. If you don't want it, I can take it over to N.B.C.'

'Sure, we want it!' said Akely. 'But we also don't want to wind up in the sneezer.'

'That's why I made up this story, like,' Bimmie put in.

187

'See, the whole thing is like a play at the theater. When the curtain goes up for the second ack, we see.'

'Never mind that crap,' Patrick Aloysius cut him off, 'maybe Bimmie's story stinks but he's got an idea. The Commission hasn't any jurisdiction over you, or any more right to interfere than they would have with that there boxing play, "Golden Boy" when it was produced. You advertise it as a skit and everybody knows that when Matilda and the Cyclone get in the ring in the studio, they're going to try to knock each other's brains out.'

The two television men exchanged looks. If it could be done the audience possibility was staggering. Akely said, 'Mr Ahearn may be right, but I'd have to have our lawyers look into it.' He took his scratch pad and began to write, murmuring, 'Ten rounds ... forty minutes ...'

'Garn!' said The Bermondsey Kid, 'Matilda ain't ever 'ad to go no ten rounds. Write down two, maybe. It usually 'appens in the second.'

Akely looked up and said, 'That would give us plenty of time for our story then.' He turned to his colleague, 'Do you think you could cook up a yarn to go with this, Ivan? You can see the possibilities ...'

Bursten doodled for a moment on his note pad, 'Brother!' he said. He was suddenly having visions of a show that would make him the most talked-of producer in the business. Then he said, 'That's not such a bad idea, you know, this baby kangaroo lost in the desert. Maybe they've got one in the zoo. For marquee value we get Sean Connery to play this fellow who comes along and finds him ...'

'I like it,' said Akely.

'It's a natural,' said Kaplan.

'If we could get Mia Farrow to do the bit as his wife later on,' Bursten added, 'she's hot right now.'

'I like it,' Akely repeated, 'it could work. I'll put it up to F.G. He's just enough of a fight nut himself to go for it.'

Bimmie said, 'Then it's in? My idea?' His narrow chest swelled with pride as he thought how proud Mr Pockhurst would be of him.

'Except for one small item,' said Bursten, deflating everyone there. 'A show like that would budget around three-quarters of a million dollars. We'd have to find ourselves quite a sponsor.'

The legal department having determined that if all those taking part in the show were listed and paid as actors, and in view of the emergence of a new kind of entertainment known as the living theater in which the performers did whatever came into their heads at the moment, gave the green light.

The next meeting at C.T.S. took place a week later in the sumptuous office of F.G. himself, the President of the Company. F.G. had a name, but so exalted was it that to use it instead of just the initials verged upon blasphemy.

All those from the first meeting were present. In addition there were three newcomers, a Mr Victor Clench of the National Advertising Agency of Blair, Howard, Clench and Morrison, accompanied by a little roly-poly butterball of a fellow by the name of Frederic Van Houven, and Whitey White, manager of Cyclone Roberts.

Mr Clench of the advertising firm, a large angry-looking man with a chip on both shoulders had a slight cast in one eye that gave him somewhat the glare of an irritated bull, took an instant dislike to Mr Herman Kaplan the agent.

F.G., precisely calipered, machine-made and turned out according to the Corporation Executives factory specifications even to the square jaw, bland, unlined face and distinguished gray at the temples, regarded Bimmie in the manner of something he would only handle with tongs.

Mr Van Houven's hair was cut *en brosse*. He had a small, petulant mouth and slightly porcine nostrils. He sat like a Buddha, only his shrewd eyes moving restlessly round the group. When they looked upon the tobacco-stained mouth of Ahearn and the scrambled features of Billy Baker, they exuded an expression of great distaste.

Whitey White, Cyclone's manager, a taciturn creature who lived up to his name by being practically an albino, looked upon no one with favor.

Yet they were all drawn together there with one aim in common, to make a killing.

The larynx which the factory had installed in F.G. opened. The thin, firm executive lips moved and he said, 'As you probably know, it is no longer a simple matter to produce a sponsor or even a series of sponsors for a sporting event which appeals largely to a male audience. The shaving people are overextended; the F.C.C. has never been willing to let us touch liquor and the cigarette companies are going off the air altogether. This is too big a spectacular for a conglomerate of sponsors. We prefer to deal directly with one important and reputable advertiser and we have been so fortunate as to engage the interest of Mr Van Houven, here, through the distinguished agency of Blair, Howard, Clench and Morrison—Mr Clench, here, representing Mr Van Houven...'

The wild-bull glare in Mr Clench's eyes had increased in intensity as he interrupted, 'But not for three-quarters of a million dollars!'

F.G.'s lips tightened and he said, 'That's one of the items on the agenda. You've a copy of the budget before you. We have shaved it where we can. Our lawyers say if we're to maintain our position that we are putting on a dramatic skit, we must use up the allotted time, no matter what the outcome of the... ahem... encounter.'

Mr Clench turned his glare upon the two managers and said, 'Three hundred thousand—a hundred and fifty thousand each—is too much to pay...'

F.G. cut in smoothly, 'There may be a little air in that figure and I was about to ask these gentlemen if they wouldn't see their way clear to taking a cut.'

Whitey White removed the toothpick upon which he had been chewing, contemplated it for a moment and then said, 'We don't have to fight no kangaroos. The Cyclone can build up to another fight with Lee Dockerty without that. We're laying our chances on the line. A hundred-and-fifty grand.'

F.G. turned to Ahearn and said, 'I see Mr White's point. As I understand it, it's really you gentlemen who need this match, would you be willing to...'

'We would not,' Patrick Aloysius concluded for him, and managed to look deeply hurt at the suggestion. 'They get a hundred-and-fifty grand; we get a hundred-and-fifty grand. We're the draw and ought to be getting sixty-forty. We've got a very sensitive proposition here. We don't take less than

190

they do, unless,' he added sardonically, 'you want to make it winner take all.'

Whitey White restored the toothpick and said, 'Since when have you been a comedian, Patrick Aloysius? A hundred-and-fifty grand, or I pick up an offer I got on the Coast.'

F.G. asked of Clench, 'What is your client's final price?'

'Half a million and not a cent more. We've been doing some budgeting on our own over at Blair, Howard, Clench and Morrison.' He turned his baleful eyes upon F.G. and said, 'And what's more, you can afford to pick up the tab for the rest. You'll blanket NBC, ABC, and CBS during prime time and get national coverage, besides what the sports writers and Duke Parkhurst write. Take it or leave it. We know you've been to all the other agencies.'

The room lapsed into a state of silent shock. No one but Clench could talk to F.G. like that, no one but a member of a firm that controlled more than a hundred million dollars-worth of advertising budgeting.

But F.G. was not head of a giant corporation for nothing and he was noted for being able to make decisions. He said smoothly to Clench, 'You're right, Bill, we can afford that. It's a deal. We'll get our lawyers onto the contracts.'

Everyone around the conference table relaxed as tension was dissipated and even some of the glare went out of Clench's eyes as he said, 'I knew you'd see it that way, F.G.' He turned to his client and said, 'Are you satisfied, Mr Van Houven?'

The butterball head nodded an affirmative but he said nothing. F.G. made the little speech he did to all sponsors when a deal was agreed upon, 'We think you will be very happy when you see the results, Mr Van Houven. You're buying not only a great attraction but an original idea and I wouldn't be surprised if we got Telstar coverage abroad as well. Your product will be on everybody's lips.'

Akely and Bursten both blew up at this point and then tried to cover hysterical laughter with fits of coughing and slapping each other on the back for relief, but F.G. never batted an eye. 'If your sales are not tripled within six months,

I've very much missed my guess. I can promise you that Chicago Television Syndicate will be behind you one hundred percent.'

Mr Van Houven nodded again.

Clench held up his hand and said, 'My client has just one small condition to make, an obvious one which I hardly need mention, still...' The buzz of satisfaction around the conference table died away and some of the earlier tension returned as they all stared at the advertising man, waiting for him to finish his speech.

'Both fighters, of course, will be wearing Ajax supporters.'

To everyone's astonishment it was Billy Baker who said, 'Both fighters will be wearing *what?*'

The Bermondsey Kid up to that moment had been practically invisible and in fact F.G. did not even know who he was until Akely leaned over and whispered to him. The President said, 'It's nothing to be upset about Mr Baker, since an athletic supporter and cup is part of every fighter's equipment and it seems to me only natural that since Mr Van Houven has so kindly agreed to sponsor this show that he should wish the fighters to make use of his product.'

Bimmie said, 'Sure, sure, anything you say, Mr F.G.'

But Baker, who had heard what sounded like an alarm bell, said, ''Ere, wait a minute! What product?'

F.G. took the floor again. He said, 'I took it for granted that all of you gentlemen were familiar with our sponsor here; you probably are with the trade names of various of his articles. Like all great captains of industry, Mr Van Houven does not advertise himself. He is president of Van Houven and Greathouse, the largest manufacturers of surgical appliances, bandages, and orthopaedic aids, one branch of which is devoted to the production of...' and here F.G. smiled lovingly upon Mr Van Houven, 'Correct me if I'm wrong, sir... The production of,' he repeated, 'the Ajax Webbing Reinforced Athletic Supporter.' Here his gaze took in the rest around the table. 'I am sure you've seen their slogan in their national advertising, "It's the Webbing that Protects You". Am I right, Mr Van Houven?'

The butterball head produced another affirmative.

'If you will look into the folder on the desk before you, you will see the manner in which we expect to present this in the commercials. We show pictures of various athletes including, of course, a boxer, with a diagramatic indication of exactly how the Ajax Supporter affords him maximum protection in the sensitive areas. If I say so myself, the mockup of the see-through or illustrated portion has been done with the utmost good taste and I feel sure could give no offense to anyone. It is in effect simply another garment. Of course, when the fighters are on the screen and actively engaged, the audience will not *see* the Ajax beneath the trunks, but by then they will *know* that it's there.'

Billy Baker said, 'Matilda don't wear no trunks; he don't wear no supporter; he don't wear nuffink!'

It seemed as though the heavy silence following upon this statement would never be broken until F.G. said, 'Doesn't wear trunks? But every man . . .'

''E ain't no man,' said Billy, ''e's a h'animal. 'E's never wore nothing but a bit of tape or rope around 'is middle to show where the foul line comes. 'Ave you ever seen the knockers on a kangaroo? They ain't located the same as like with the rest of us. You tie one of them things around 'im and you might as well 'andcuff 'im as well.'

F.G. filtered the information through his executive mind and quickly came up with the friendly solution. He said, 'Well, I'm sure that in the case of Matilda Mr Van Houven would waive his condition since The Cyclone will be wearing . . .'

For the first time in the entire session a word emerged from the little polyp mouth. It was, 'No!'

Mr Clench rekindled the glare in his eyes. 'You heard what my client said, F.G.'

At this point Ivan Bursten spoke up and said, 'You've got another problem there to lick, F.G. This picture will be going into millions of homes in the Midwest. I've taken a look at this animal. Brother, they're the size of oranges! You show those on the screen and you'll choke up your switchboard the next three days with complaints from the Church, the Legion of Decency and all the rest of the do-gooders. Permissiveness hasn't got that far with us yet.'

Billy Baker asked, 'Do you have agricultural shows on the telly?'

'Of course,' Bursten replied. 'What's that got to do with it?'

'Do you put pants on the bloody bulls? I've been showing Matilda for more than eight years now, and never had no complaints. I've never trained Matilda to wear nuffink around his legs and I aren't going to start now.'

F.G. said, 'I wouldn't worry about that question too much, Ivan, you can angle the shots away from that area.' He then addressed himself to Van Houven and said, 'Don't you think, Mr Van Houven, it will be even more effective if just Cyclone Roberts is wearing the Ajax? We can reword the commercial to emphasize that one fighter is Ajax-protected and the other isn't and sloganize, "Would you take a chance like this with your safety?"'

The small mouth parted again to let escape another 'No!'

Mr Clench turned and leaned down to his client and they conferred in whispers for a moment. The advertising man then said, 'My client's purpose in sponsoring this show is to familiarize the public with the nature and value of his product. With both fighters wearing the Ajax, he's got full coverage. With only one protected in this manner, it leaves some doubt in the viewer's minds as to the value of the safeguard afforded. The impact is only one half of what it would be with both men equipped. My client is spending half a million dollars to put his message across. Besides which, as Mr Bursten has pointed out, there is the moral problem of those appendages. Both contestants must wear the Ajax, or we withdraw our offer.'

'My Matilda don't go into no ring wearing any fur-lined jock strap!' Baker declared.

'Our Matilda, if it comes to that,' put in Patrick Aloysius, allowing his sardonic gaze to rest upon the advertising managers, 'is a lovin' kind of fella and I ain't ever seen him strike a blow in anger. But you tie one of them Ajax things around his arse and we wouldn't want to be held responsible for the charge; manslaughter probably, or homicide while

194

temporarily of unsound mind. You don't want to get our Matilda mad, Mr Clench.'

The advertising man and Van Houven conferred in undertones again but it was obvious that the manufacturer had a one-way head which could only wig-wag 'No', causing Akely to whisper to Bursten, 'Absolutely classic! He ought to be preserved in alcohol or stuffed. The one-hundred percent perfect example of the sponsor.'

Clench knew when to stop. He delivered his ultimatum, 'Either he wears our supporter, or we don't sponsor.'

Billy Baker delivered the counter ultimatum, 'You make him put on your bloody hammock and we don't fight!'

'Wait!' The exclamation came from none other than Bimmie who had been growing progressively paler and his hair rising in successive layers during the course of the altercation. 'Mr Clench, Mr Van Houven, I understand you, like, now you're both businessmen and I'm a businessman too. Like Herman Kaplan here will tell you, I've been booking acks now for five years and anybody who books with Bimmie gets a square deal. Book with Bimmie and you'll be satisfied. If a gentleman comes to me and says he wants a ack where an elephant can sit up on his tail and sing, "Mother Machree", that's what he'll get, if there's an elephant anywhere can do this. Mr Clench, Mr Van Houven, half a million dollars talks loud enough so I can hear. Mr Clench and Mr Van Houven, you're both clients who should be satisfied. You say Matilda should come into the ring wearing the Ajax Webbing Reinforced Athletic Supporter, then I say Matilda comes into the ring wearing the Ajax Reinforced Athletic Supporter, and for that price, as far as I am concerned, you're entitled to paint on his . . . on the side of the product how much it costs and where you can buy one. If you got any outsize, jumbo, number fifteens in stock, you send one along to us and that's what Matilda's wearing when he comes into the ring.'

The Bermondsey Kid was up in arms, 'Now look 'ere,' he shouted, 'who says so? You 'eard what I said. You ain't puttin' any bleedin' nappies on my Matilda! Who owns this kangaroo?'

'We do,' Patrick Aloysius said quietly, for he was the only

one who had seen the almost imperceptible droop of Bimmie's left eyelid. 'We own him, Billy—you, Bimmie and I. But you're out-voted, Billy boy, two to one. Bimmie's right and I go along with him. When you're on the big time you can't worry about a lot of balls—sorry, gentlemen, I mean trifles. Like Bimmie here says, we accept the condition of Mr Van Houven. If you can fit him up with it, we'll see that he wears it into the ring and you can write that into the contract. Is that satisfactory?'

Mr Van Houven's small speaking orifice opened to release an unfamiliar word. He said, 'Yes.'

'Great!' said F.G. 'Mr Bimstein and Mr Ahearn, you're true sportsmen and I'm glad that I've enabled you to see it my way.' For he was convinced, of course, that as usual he had effected a compromise and won the day. 'Gentlemen, the deal is on.'

# XIX

'THE MATILDA STORY' as produced and broadcast by the Chicago Television Syndicate network at 9.0 p.m. the night of November 16th starring Sean Connery as Jack Hardy, Mia Farrow as his wife and Matilda as himself had something for everyone—Dad, Mum and the kids. From the moment that Sean Connery as the lonely swagman, or tramp, was seen trudging through the sandy scrub of the Australian desert, Mr and Mrs Average American family never stirred from in front of their television box, black and white or in color. Bars did a roaring trade. The TV lounges of hotels, clubs and boarding houses were jammed. Cars equipped with television parked at the roadside. Aircraft in flight put on the show in place of a movie. Two spacemen in orbit got it on their set as a welcome surcease from the eternal orders from Houston. Wherever the broadcast waves reached, wherever there was a set that would receive them, there 'The Matilda Story' glittered and flickered from the surface of the tube.

The children screamed with delight when Jack Hardy came upon the young kangaroo lost in the wilderness and Sean Connery played it to the hilt for sympathy and pathos, thus immediately engaging the affections of the women.

'Why, hello there, little fella,' said Jack Hardy in Connery's rich, deep voice, 'what are you doing out here all by yourself and where's your mum? I'll bet you're lost. You don't seem as though you'd had a square meal for days. And look at that funny blaze on your forehead! Ain't ever seen a joey with a blaze like that before.'

When he picked up the cuddly, furry animal and cradled him in his arms, all the kids went, 'Oooh!' and 'Aah! Ain't he sweet!' and the maternal instincts of all the mothers were stirred. The dads were content to wait for what had been promised them.

'We'll just get the old billy-can going,' said Jack Hardy, 'and warm up some milk for you.' The relationship between the man and beast became really touching as he heated his can over the camp fire and fed the kangaroo to an obbligato of contented squeaks.

'There you are, Joey' (that being the nickname for all baby kangaroos in Australia), said Jack Hardy, 'and now we'd better go and try to find your mum before it gets dark.' Still cuddling the kangaroo who was now licking his face in gratitude, the wanderer marched across ten million screens until he came to a eucalyptus grove. There in close-up they saw a large, female kangaroo standing erect with a most concerned expression on her face. 'There you are, old boy. There's your mum! Off you go!'

But the young kangaroo hesitated and seemed reluctant to leave his benefactor.

'Want to come with me, old fella?' said Jack Hardy, 'You'll be better off with your mother. I've a fair piece to go before I can ever settle down. So long, then, Joey! Good luck!' and the lonely swagman went trudging off into the twilight. Everybody loved that, even the men. If never before, television was justifying its existence.

The first commercial came on, a football player armed *cap-à-pie*, who announced 'I feel safe with my Ajax Webbing Reinforced Athletic Supporter,' while the camera double exposed the area involved just sufficiently to give a hint of the article. The story resumed with the legend on the screen, 'TEN YEARS LATER'.

Nobody wondered or much cared how, during the interim, Jack Hardy had become a once prosperous boxing manager now in difficulties and married to a loving and faithful wife. Miss Farrow's looks and figure kept the men occupied, while her sympathetic portrayal of the loyal helpmate standing by her man facing bankruptcy unless an opponent could be found to fight the Australian champion, engaged the wives.

'Have faith,' Mrs. Hardy encouraged her husband, 'something will happen to save us. You've been a good and kind man all your life.'

'Christ, what corn!' Akely had remarked at this part of the script.

'Anything to keep the cops out of the joint,' Bursten had said, as he wrote the next scene which moved women, children and all the drunks in all the bars to tears.

'They saw Jack Hardy desolate and wandering the streets of Sydney. On the outskirts a circus was playing and a huge twelve sheet advertised, 'CAPTAIN BILLY BAKER AND HIS CHAMPION BOXING KANGAROO, MATILDA.'

Stirred by memories of his early days in the bush, Jack Hardy went in and back stage encountered Matilda.

For a moment man and kangaroo confronted one another. The next instant Matilda was embracing Jack Hardy and covering his face with kisses. (When it came to love scenes, Matilda was a natural actor. Mr. Connery later declared that he nearly drowned.)

'He knows you!' cried Captain Billy Baker.

'It's my little Joey,' said Jack Hardy, 'whose life I saved in the desert. He's grown up but I'd know him anywhere by that blaze on his forehead. Maybe he'd pay me back by fighting the middleweight champion of Australia for me.'

''E'd do anything for you, sir,' said Captain Baker. 'You sived 'is bloomin' life.'

The commercial now showed a baseball player wearing the shadowy supporter. In seven million of the ten million homes into which this entertainment was being beamed, eight-hundred-and-ninety-thousand little girls asked, 'Daddy, do you wear one of those?'

Ivan Bursten wasted no more time and the very next scene showed the huge bill poster advertising the fight between the middleweight champion versus his challenger Matilda, Australia's champion boxing kangaroo—ten rounds to a decision. This was where the men sat up. Their part of the show was about to begin.

For by the magic of the producer's art the watchers were

now transported to a replica of an indoor sports arena set up for fight night with a regulation, twenty-foot ring. Working Press, telegraphers, spectators, officials and all, caught in that moment of expectancy before the participants in the championship appear for the battle.

By now watchers in Chicago, as well as other cities of the Midwest, had begun to identify members of the Working Press at the ringside, including the huge figure of Duke Parkhurst.

When the cast of characters would be reeled off at the conclusion of the broadcast, they would all be given fictitious names to conform with the gimmick, but in the meantime they were there. The fact that the fellows playing their telegraph operators were real telegraph operators and the bugs they worked, the clicking of which provided background out of the TV speakers, were connected to wires that went live into the various sports departments was nobody's business, any more than that the chap playing the referee was a well-known arbiter from Detroit and the judges and timekeepers from Cincinnati, F.G. has spared no expense with his cast.

The moment had arrived. The commercial, the one showing the boxer, had been aired and then the first of the contestants with his entourage of seconds and handlers entered the ring. The legend 'CHAMPION' was emblazoned on the back of his bathrobe, but everybody recognized the scarred face of the veteran Cyclone Roberts, while those with color sets had the additional identification of his carroty-red hair, not to mention the albino-blond Whitey White in his corner. The telegraph instruments chattered more loudly. The actors who comprised the visible crowd in front of the painted backdrop of those in the rear seats, began to buzz and hum.

And then a gasp went up not only from the paid hands in the studio who had not been told what was coming but also from the millions of viewers. For what their screens were showing them was Matilda, the famous boxing kangaroo and sharer in the middleweight championship of the world, not hopping in under his own steam to leap over the top rope

200

as everyone had read was his custom, but *carried in* by his three handlers.

The three men gingerly hoisted him through the ropes and onto his stool, when it was observed that his nether parts appeared to be wrapped in some gigantic type of diaper extending to well below his knees. Even on the smallest flickering television screen of the smallest portable it was possible to see that the expression on Matilda's face was frantic.

Ivan Bursten himself was at the mixer in the control room, pushing the buttons that designated which of the scenes showing on the twelve monitors of the carefully placed cameras would appear. He lingered dramatically on the anguish reflected in every lineament of the kangaroo's elongated muzzle and bewildered eyes.

An almost universal gasp of horror went up from the viewers including the women, who by this time had been brainwashed into realizing that beneath the diaper Matilda was hobbled by the Ajax Webbing Reinforced Athletic Supporter. Strong men swore helplessly in the presence of their children. Women bit their fingernails.

Emotion ran even higher, building up into almost unbearable tension as the sets showed Matilda coming forward to receive his instructions from the referee in pitifully, hopelessly awkward hops which not even his powerful tail, emerging from the contraption, could alleviate.

A close-up showed Matilda's expression to be still more desperate; his brow furrowed, the whites of his eyes showing as they rolled, muzzle drawn back from his teeth and it was all his handlers could do to soothe him, rubbing his back, fondling his ears and whispering to him.

Never had so many men in the United States of America used the words, 'son-of-a-bitch' so simultaneously. Never had so many women cried out aloud against cruelty and inhumanity, since it was obvious that it was going to be impossible for Matilda to fight his fight or, for that matter, any fight under those conditions. They had forgotten by this time the crucial situation of Sean Connery and Mia Farrow who must have victory to survive. Their emotions were wholly engaged by Matilda.

And yet when sensation came to top sensation, nobody was prepared for it.

When the bell rang, triggering the reaction that sent Matilda leaping up off his stool, as if by magic the wrappings and the apparatus it contained—the number fifteen-size Ajax Webbing Reinforced Athletic Supporter—was whipped from his legs to disappear in his corner and the kangaroo, now free and nude except for his own fur casing, reached the center of the ring in one hop and in battle pose.

Before the mass sigh of relief had time to dissipate into the stratosphere the Cyclone, who as his name indicated was a swarmer, was all over Matilda and the kangaroo out of sheer necessity was compelled to give one of his supreme exhibitions of defense, ducking, slipping, shifting, picking off punches left and right, to come unscathed through the hurricane of blows. So marvellous was the action during these first two minutes that the extras in the studio began to cheer.

Cheers were echoed from homes throughout the seven States of Chicagoland. Yells arose from bars and cars, motels, hotels and clubs, wherever the silvered picture tube relayed the drama.

Yet there was still a topper to come. Suddenly a round little man as broad as he was tall, with a cropped, butterball of a head, leaped from his ringside seat waving a paper and shouting something which would have been unintelligible except for an especially sensitive microphone.

'Stop! Stop the fight! He's got to wear the Ajax! It's in the contract!'

The clever Bursten immediately split his screen so that on one half was Matilda giving his glorious exhibition of the Manly Art and on the other, the irate little man jumping up and down like Rumpelstiltskin.

Now he was brandishing his paper under the nose of one of Matilda's seconds, later identified as Patrick Aloysius Ahearn and the close-up showed the old manager saying something. A lip reader would have had no difficulty in translating what Ahearn was telling the fat man to do with himself and at this point Bursten switched back to full screen and the action in the ring.

Here the Cyclone paused to catch his breath and let his eye be distracted for a fraction of a second by the altercation going on in Matilda's corner. This was all that was needed.

While Matilda had been thoroughly enjoying frustrating his opponent's violent attack, a beautiful foil to his own accomplishments, his nerves as well as faith in his friends had been temporarily shaken by what they had done to his legs, even though at the bell he had gained his freedom. Under other circumstances he might have enjoyed one or two more sessions with the Cyclone. As it was, with an upsetting beginning such as he had been compelled to endure, enough was enough. He did not so much as trouble to feint, but whipped over such a lightning right that its impact on the side of the Cyclone's jaw and the Cyclone hitting the deck were practically simultaneous.. The referee did not even bother to count.

The viewers, mothers, sons, daughters, husbands, wives—were weeping in one another's arms.

Matilda having confirmed Billy Baker's prophecy that the fight portion of the drama was not going to take up a great deal of time, it now became necessary to screen one of Mr. Bursten's two alternative endings, namely the victory celebration in the home of Jack and Mrs. Hardy in which Matilda figured largely, photographed from all angles stuffing himself with his usual sweets, as well as lapping up several quarts of ice cream out of cartons.

Celebrities of stage and films who happened to be in Chicago at the time had been persuaded to appear and play themselves at this party, in which the lines and action were largely *ad lib,* until as time ran out the last scene irised out on Mr Connery and Miss Farrow with Matilda between them, the latter bestowing kisses with equal affection and moisture on both. Music up and the credits began to unroll.

Fifteen million people, since the Nielson ratings figured five persons to a set that night, went to bed wholly satisfied as well as forever committed to the Ajax Supporter.

Bursten, of course, had had a second ending prepared should Matilda have been so unfortunate as to suffer defeat at the hands of the Cyclone. It would have been a Chaplinesque finale where Jack Hardy joined Matilda's act as a clown, and the drama concluded with a fade out as the circus wagons silhouetted their way over the horizon into the rays of the setting sun.

Bursten almost regretted not having been able to use it, it was so beautiful.

The real, and strictly private celebration, took place somewhat later in the Pump Room of the Hotel Ambassador West, that famous Chicago landmark decorated in the style of Beau Nash's era in Regency Bath, complete with waiters in livery and coal-black coffee boy togged out Ottoman style.

Swank was the word for the place. Here Chicago Society came for late suppers consisting chiefly of *brochettes* served on flaming swords and where Patrick Aloysius had already endeared himself by ordering scrambled eggs and producing that old chestnut, remarked, 'Let's see you bring that on one of your goddam flaming swords!'

Still, Duke Parkhurst, who was recognized by all, lent some tone to the gathering numbering Birdie McBride, Bimmie, Ahearn and Billy Baker.

As always, even though during the entire evening she spoke hardly a word but sat silently clinging to Parkhurst's arm or hand, Birdie cast her spell over all of them with the possible exception of Patrick Aloysius whose half sardonic, half amused expression concealed a worry. The upset that Ahearn was experiencing did not stem from her looks, to which he was as susceptible as the rest, but from the fact that she was there and there could be no doubt that she and the Duke were gloriously in love.

Parkhurst, however, was totally unaware of what was going on inside the head of the veteran fight manager. Secure in the affection of the girl at his side and soothed beyond words by the fact that she never interrupted, he was engaged in quizzing Bimmie, 'Come on, you little bastard,' he coaxed, 'How did you work it?'

'Well now, you see,' Bimmie replied. 'Mr Pockhurst it was just a gimmick. We gimmicked the Ajax for a breakaway like now, see, this Apache team Claude and Claudette that I booked into the Sixteen Club once. They had a ack where she's wearing a sort of tight skirt when it starts out, only it's a break-away, see? So after the guy throws her around a bit, he suddenly tears off her skirt and she got nothing but a little pair of pants underneath. But the skirt has to be gimmicked so it comes right away and don't spoil the ack by Claude having to yank at it, trying to pull it off her. So you see, Mr Pockhurst, we took the Ajax down to some pals of Herman here, in the costuming business and they gimmicked it for us so that when the bell rang and Matilda gets up off his stool, Mr Ahearn here got this finger on it and it comes away.'

'And just in time, if you asks me!' put in The Bermondsey Kid. 'That 'roo would have had a nervous breakdown if he'd 'ad to wear that bloody chastity belt another minute.'

Parkhurst said, 'I still don't get the picture. What was biting the little man running around the ring yelling to stop the fight, and waving that piece of paper?'

'Our sponsor,' said Patrick Aloysius, letting the two words dribble out of the side of his mouth.

Birdie gave him an angelic smile to compensate for the fact that she did not know what a sponsor was.

'See, Mr Pockhurst,' Bimmie tried to explain, 'it was the contrack. But he didn't read it right. Like now, it only said Matilda would be wearing his jock strap when he come into the ring. See, he said unless both fighters was wearing the Ajax Webbing Reinforced Athletic Supporter, he wouldn't ack as sponsor and we couldn't have the fight. So we put it in the contract like he said, that Matilda would be wearing his product when he come into the ring, but we didn't say he'd have it on him after the bell rang.' And then he added quickly, 'If you're in business, you're supposed to read a contract, ain't you, Mr. Pockhurst?'

He voiced this for a vision of Hannah and her, 'Is it honest, Bimmie?' had suddenly come up and he was anxious that Mr Pockhurst should not think him less than such. Whenever he indulged in anything that might be construed as even slightly shady, there was Hannah querying him with tears behind her luminous eyes. The affection displayed during the evening between Mr Parkhurst and Miss McBride only served as an additional saddening reminder to Bimmie that somehow he had managed to love and lose.

The big columnist, however, only threw back his head and roared with laughter and said, 'Oh brother, tomorrow's piece, and tomorrow and tomorrow!'

The little tremulous quiver formed itself at the corner of Birdie's mouth and she hugged Parkhurst's arm. She did not know what he was laughing about but was just happy that he was happy.

Bimmie was dying to ask Parkhurst about the possibility of the fight with Dockerty for the million dollars but he remembered that he had been pledged to secrecy. But after midnight when they had finished eating, he could contain himself no longer. He said, 'Mr Pockhurst, how about it now? Remember what you said before like, about you know ...' and here he gave him a violent wink. 'But I kept my mouth shut like you said, even with Paddy Aloysius and Billy here, who don't know see, about the million dollars.'

Parkhurst nodded. He said, 'Okay. We still want to keep this in the family, but I want you all to know about it now. Are you three fellows willing to take a gamble?'

Bimmie said, 'For a million dollars, who wouldn't?'

Ahearn asked, 'What are the odds?'

Parkhurst said, 'You figure them out. The *Daily Mercury* is prepared to promote the Matilda-Dockerty fight for the middleweight championship of the world for its Free Food Fund for Hungry Children, in Jericho Stadium next summer. We'll get behind it with everything we've got, and you know what our circulation is. The gate will scale at something a little over three million dollars.'

Even Patrick Aloysius, who was shocked at nothing, was taken aback by this staggering sum. 'Brother,' he said, 'but can you get Dockerty?'

'If you'll do what I say.'

Bimmie's hair had started on its rise, layer by layer, illuminated by another flaming sword going by. He said, 'Mr Pockhurst, you say it—you got it.' The elevation of his locks was caused by the turmoil beneath, as he was figuring what fifty percent of the purse would come to if they could not get sixty. But then remembering, he asked, 'But what's the gamble, Mr Pockhurst?'

'You don't get paid,' the Duke replied succinctly. 'You

donate your entire end to the Fund.'

The hush that fell upon the table in its bay was louder even than the clatter of dishes and the noise of laughter and conversation going on at either side.

Finally Bimmie found his voice, 'But Mr Pockhurst, the million dollars, like now you and Mr Clay said we would make when you told me...'

'You'll get it—after Matilda has knocked out Dockerty and is recognized incontestably as the middleweight champion of the world. That's the gamble. After that you can clean up. You can figure on half a million for exhibitions and a million for his first defense of the title. Are you game?'

Parkhurst had expected that his suggestion would come as something of a shock to the trio but had not been prepared for the dismay reflected on the faces of all three, the sudden silence that fell upon them once more and the glances they exchanged.

Patrick Aloysius said, 'You mean we fight for nothing?' As an old-time fight manager it was the idea that was so abhorrent to him, even though he already had a glimmering of what Parkhurst had in mind.

Parkhurst said, 'Not a penny. Matilda can give some exhibitions with Billy Baker between now and then and you can keep what you take in during training, which will be a nice piece of change. But can't you see it's the only way we'll ever get Dockerty into the ring again with Matilda. When we announce that we've offered Dockerty and Pinky Schwab the champion's end of the gate and you're prepared to give the whole end of your purse to the charity, they'll have to agree or never fight again. Every State in the Union will bar them.'

Patrick Aloysius heaved a great sigh and said, 'The Duke is right, boys. Until Matilda gets Dockerty back into the ring and takes him in a regularly sanctioned, fifteen-round championship bout, he won't be recognized as world's champion, no matter who he licks. Let's not kid ourselves. It all started as a gag but there's a time when the laughing's got to stop. It's okay with me.'

The other two were looking at him in amazement.

Billy Baker said, 'Me pub!'

Ahearn said, 'You've got enough already for three pubs. If Matilda can take Dockerty, you can buy up a couple of breweries as well.' He turned to Bimmie and it was the old cynical Ahearn again who said, 'What Mr Pockhurst wants, Mr Pockhurst gets, eh, Bimmie?'

'Sure, sure!' said Bimmie. 'If it's okay with the others, it's okay by me. I make it uniminous.'

But his hair had not yet subsided and Parkhurst felt that he did not sound entirely convinced. Still, they were in accord and then Billy Baker added, 'Right you are, guv'nor, Matilda took him once, he can take him again.'

Parkhurst was satisfied that the three would not go back on their agreement and that under those conditions it would be impossible for Dockerty and Pinky to side-step them, or indulge in those stalling and delaying tactics of champions trying to duck a return match. Nevertheless he was aware that some of the glow had gone out of the evening but put it down to the natural reluctance of anybody to do something for nothing, particularly in the boxing business.

Bimmie asked, 'What about a licence for the fight in New York, Mr Pockhurst? See, the Boxing Commission here wouldn't let us have the fight and we were practically run out on our ears for asking. If they won't give us a licence in New York...'

Parkhurst said, 'New York is not Chicago.'

Paddy Ahearn muttered, 'New York is worse. You've got that ape Wild Bill Wildman.' He seemed sunk in the deepest pessimism. 'He's got it in for me for chiseling those papers for Bimmie and The Kid, here, before he knew what it was all about.'

Parkhurst nodded and said, 'I know. But I've an idea or two which may make it difficult for him to refuse. The main thing is for all of you here not to shoot off your mouths, start demanding things and getting his back up. When we announce our readiness to promote the fight for the Fund and reveal that you'll be working for free, the public will demand the match. When the fans yell loud enough and show enough muscle, that's when we move in on Mr Wildman.'

He glanced at his watch and said, 'Thanks for the party.

Time for some shut-eye. Up we go, Birdie. See you tomorrow, boys. We're staying here at the hotel if you need to get in touch with me.'

Left to themselves the three heavily contemplated their half-empty glasses, but the gloom seemed to sit the thickest upon Patrick Aloysius. Finally The Bermondsey Kid said to Ahearn, 'Oi, mate, it ain't all that bad, is it?'

'Worse,' Paddy Ahearn said morosely.

Bimmie asked anxiously, 'You mean we shouldn't have agreed to go in like we said? And losing out on a million dollars? I thought you were off your nut like, when you said okay.'

Ahearn shook his head. He said, 'No, that's all right. Sometimes you got to take a chance, even a big one, and he knows what he's talking about. How were we going to say no? What's bad is that Parkhurst's got some trouble coming up he doesn't know about—him and the broad with him.'

'The bird?' said Baker. 'Cor, isn't she the little beauty?'

Bimmie said, 'Didn't she used to be in a show on Broadway? Boy, she's a looker!'

Patrick Aloysius, his jowls seeming to sink deeper in his chest, said, 'Yes, that's it—the bird, as you Limeys call them. Birdie McBride and Duke Parkhurst; would you believe it? The smarter they are, the harder they fall. A wiseguy like Parkhurst!'

Bimmie asked, 'Why, what's wrong with Birdie and Mr Pockhurst going for her? They was nuts about each other, just like me and Hannah was.'

'Nothing,' Patrick Aloysius Ahearn concluded glumly, 'except I remember the talk along Broadway is that whoever Uncle Nono is, used to be thick with Birdie at one time. And if he's using her for bait, our guy's taken it, hook, line and sinker and the goddam pole along with it.'

# XX

THE CYCLE OF seasons rotated once more until baseball, the harbinger of summer, filled the sports pages and all the time pressure for a return match between Lee Dockerty and Matilda had been mounting.

Following upon the revelation by the *Daily Mercury* that Matilda's management had agreed to contribute their entire purse to the Free Food Fund for Hungry Children, Walter Mason, President of the Jericho and Metropolitan Stadiums, had consented to donate his giant Long Island amphitheater for the same cause, while the full amount of sixty percent, the champion's percentage, had been offered to Lee Dockerty and company.

Put on the spot, Pinky was squirming. He would not say yes and he would not say no, but kept retreating behind every dodge known to a fight manager to avoid combat.

Letters to the editor in favor of the fight began choking the mail bags not only of the *Mercury* but other New York newspapers. Editorials began to appear in out-of-town journals. Hippies United came out for Matilda. Mass meetings of protest against Dockerty's intransigence were held and a march-past by the building housing the offices of Boxing Licence Commissioner Wild Bill Wildman saw placards raised aloft: 'DOCKERTY IS A YELLOW FOUR-LETTER WORD', 'SEND MATILDA TO CONGRESS', 'LOVE MATILDA. HATE DOCKERTY', 'KANGAROOS SHALL INHERIT THE EARTH', and 'THE SYSTEM IS AGAINST MATILDA: DESTROY THE SYSTEM!'

During the procession seven stink bombs were exploded before the entrance to the building, not exactly advancing the cause since a reporter wrote that the odor arising therefrom was indistinguishable from that of those who had set them off.

A Congressman from the district where the *Mercury* had its office and who was known for his anti-racist feelings, spoke up from the floor of the House to ask whether justice and non-discrimination was not intended according to the Constitution for animals as well as men. Borne aloft by the gas of such publicity Bimmie, Patrick Aloysius and Billy Baker were raking in coin for daily exhibitions given in Donohue's Gymnasium under the now personal blessing of The Professor himself.

During this period a permanent home had been arranged for Matilda in Ryan's Stables, where several stalls had been knocked to form a luxurious suite which when occupied by Matilda was under permanent twenty-four-hour guard by ex-thugs and strong-arm cops from a private security agency.

A violent and the harshest yet editorial blast from one of the most influential newspapers in the United States, *The Baltimore Sun,* was demanding angrily that Dockerty either agree to fight Matilda, or be made to surrender his title. It led Parkhurst to the pleased conclusion that it would not be long now. And that morning he taxied down to Tenth Avenue and 15th Street where Baker, Bimmie, or Patrick Aloysius, or all three were usually to be found.

The blue-uniformed special police guard equipped with gun and cartridge bandolier said, 'They just stepped out for a minute. They said they'd be back.'

Parkhurst asked, 'How is he?'

'Who—him?' replied the guard. 'He's all right, he don't never give no trouble.'

Parkhurst went to the stall and peered through the wire netting which had been carried up as far as the roof. Not that Matilda ever would have thought of departing the premises; he was thoroughly satisfied with life.

'Who the hell do you think you are,' Duke Parkhurst queried, looking down upon the reclining Matilda, 'the Naked Maja?' For the kangaroo had assumed one of his favorite attitudes in the soft hay of his couch, that of Goya's famous model.

Matilda's poses when relaxed were a source of never ending delight and fascination to the writer. Besides the one on his back, he had another in which he lay over on his side, resting on one elbow like an Impressionist nude and always on his face an expression, *some* kind of expression. It might be introspective, reflective, curiosity, or one of nervous alertness with one ear cocked forward and the other turned backwards to intercept signals from both directions. He also had others, quiet self-satisfaction; rather an irritating quizzical one in which he gave Parkhurst the feeling that he had been analysed and found wanting; and then of course, the deliciously soppy one when he was happy and full of the milk of kangaroo kindness and only wanted to demonstrate affection.

'You funny beast,' Parkhurst said and then apostrophizing the animal, 'Matilda, Matilda, oh wherefore art thou, Matilda?

'And, above all,' Parkhurst continued, looking down upon the curious construction of the animal at ease and luxuriating, 'what are you? Who made you and why? Why should you, the most ridiculously put-together animal in the zoo, have been given this strange talent? Has the spirit of Old Bob Fitzsimmons entered your body? Are you a living proof of the reincarnation of Old Bob? You know he was once middleweight champion of Australia way back in the 1880s, before he came to America? They said he could go fifteen rounds without a glove being laid on him. And once, back in 1881, he knocked out five men in one night. Are you really Old Ruby Robert come back to us, Matilda?'

Matilda, hearing his name again, turned over on his side, leaning on one elbow, his long hindlegs stretched out and crossed, contemplating Parkhurst. This time his expression was what the sportswriter would have described as inscrutable, as though ruminating over the theory.

Parkhurst said, 'No wonder Cook had a fit when he first saw you, and his artist Joseph Banks wrote, "What to liken him to, I cannot tell, nothing certainly that I have seen at all resembles him."' Parkhurst had been reading up on the history of the kangaroo and Lieutenant Cook's voyage in the H.M.S. *Endeavour* from England in 1768, when for the first time one was shot, skinned and sketched, and an accurate picture of this impossible beast brought back to the western world.

'But why you, Matilda?' Parkhurst asked, leaning upon the gate of the stall. 'There have been other kangaroos who have boxed in circuses and fairs and carnivals, or slapped with their forelegs, but never before one like you who, like old Fitz in his prime, could make a clown out of any boxer that ever lived, right up into the heavyweight class. Were you sent to make fools of us all and show up this rotten, filthy, absurd behaviorism we call boxing and dignify by the name of sport? Is this your purpose, you silly-looking bastard, to make us all look even sillier than you do? Including me, who writes reams about the wonders of two men who put their hands into padded bags and beat one another about the head and body; their skill, their heroism, the science and beauty of it all, the wonderful manly art of self-defense.'

Here Parkhurst laughed and interpolated, 'Self-defense! Brother, that's a good one in this day and age! ... And in three minutes of sparring in a ring, you show them up as dull-witted, helpless impotents unable to defend themselves, much less land a solid punch on you. And if Billy Baker hadn't trained you to keep your hindlegs to yourself, they'd be dead as well.

'What do you think, Matilda, or *do* you think when the bell rings and a white man or black comes at you, waving red leather gloves? Is it just another kangaroo to you, or are you a machine programed to deal destruction like so many of those hard-eyed, mindless numbskulls of the game? Am I boring you, Matilda?' Parkhurst suddenly queried.

The animal rolled over again and lay this time on his stomach, his forepaws with their five, powerful talons crossed before him and looked up into the columnist's face.

'Or, maybe you've got a sense of humor and are going along with what started as a gag and are just laughing your silly head off at all of us.'

But then Parkhurst said, almost with exasperation, 'But you love it, Matilda, goddam it, I know you do! When you're up there in the ring you thrive on it. I know when someone—man or beast, is enjoying himself. And yet you're not like those crazy kids who, when they get a man going, like to torture him, cut him up and make him suffer before they finish him off. By God and Billy Baker, you've been endowed with skills and you use them to solve the problem. You've got that Mona Lisa look on your face now, but not in the ring, brother! I've watched you—concentration, fascination, study, calculation, measuring and a flash of pure, absolute, basic, sheer pleasure when you've found the opening for the final zap, just like when you're about to tackle a jumbo-size, double-strength, almond-studded Hershey Bar.'

Whatever Matilda's mental limitations were outside of the squared circle, there was one word that rang a bell and that was 'Hershey'. At the delectable sound he arose, propped himself up on his tail and hindlegs, reached up with his paws to Parkhurst's shoulders, his brown eyes full of pleading and gave him a kiss through the mesh of the wire.

Parkhurst said, 'Oh, all right, Matilda. I get it,' produced the article from his pocket, stripped it of its wrapping and handed it over. The animal took it immediately, turned around and went into his feeding crouch.

Parkhurst gazed down upon the broad, gray back and the thick tail, tree-like at its base, from whence all that tremendous punching power must stem, just as all the great hitters of the ring had always backed their blows by shifting their weight and getting their bodies behind them. He smiled and said, 'You and I haven't got any forrarder, have we, Matilda? Well, as Bimmie would say, enjoy it in good health.'

'As Bimmie would say, what?' that individual said as he appeared in the doorway of the stable with Billy Baker.

Parkhurst said, 'Matilda was just explaining some of the finer points of the game to me and I was trying to find out whether he had any weakness.'

It was Billy Baker who replied with somewhat unusual sharpness for him, 'Weakness? Him? Have you ever seen any? He never had any that I couldn't cure. Have you seen the man yet what could stand up against him?'

'I know,' Parkhurst replied, 'and if I hadn't seen it with my own eyes, I still wouldn't credit it. Man and boy, I've been looking at boxers and champions for twenty years and still find it hard to believe there can be such a thing as a perfect fighting machine. I just dropped by to show you this editorial. When *The Sun* papers take the trouble to yell, people listen. It won't be long now. Just keep him in shape.'

After he had gone, Bimmie asked anxiously of Baker, 'Has he found out? Do you think he knows?'

Baker asked of the special cop, 'What was he talking to him about?'

'Some kind of crap about him being the incarceration of someone called Bob Fitzsimmons or Ruby Robert, who once knocked out five men in one night. I've heard about him.'

Baker said to Bimmie, 'Don't worry. As long as 'e thinks Matilda's as good as Old Bob, we're okay.'

On his return to the office, Parkhurst to his delight found that his prophecy had been borne out even sooner than he had expected, for his boxing writer O'Farrell was waving a piece of A.P. copy off the teleprinter. It read: 'New York, N.Y., 5th June. Pinky Schwab, manager of middleweight champion Lee Dockerty, said here today that if Colonel Wildman, Licensing Commissioner of Boxing of New York State, would issue a licence for the bout, he would sign for a return match with Matilda over the fifteen-round route. More later...'

Jubilantly waving the slip of copy, Parkhurst invaded the office of his managing editor and said, 'Here it is, Justus! We've cracked Pinky. We're in!' He laid the slip on the desk.

Clay's reaction justified his reputation as the office pessimist. He read the despatch through and said, 'See that nice, big little "If" there? *If* Wildman will issue a licence... They know damn well he won't.'

But Parkhurst refused to be discouraged and said, 'Okay, little ray of sunshine, but we have to try, don't we? I've got all the dope together. Are you coming with me?'

Clay said wearily, 'I suppose so. But just looking at Wildman depresses me for a month.'

Licence Commissioner Wild Bill Wildman sat tilted back

in his swivel chair, his feet clad in cowboy boots up on the desk, his vest open to show his cowboy belt, his toupé slightly askew and said, 'Absolutely no, boys. I'm afraid my answer has to be absolutely and unequivocally no. I would like to help you boys out. It's a great charity you're running but I could never sanction a fifteen-round championship fight for the middleweight title between a man and a kangaroo, now.'

The boys to whom Wild Bill was referring were Duke Parkhurst and Justus Clay.

'I see, Bill,' Parkhurst said, as though in resignation, 'well, if that's your answer, that's it. I'll only ask you then to give me your reasons and we'll be on our way.'

Wild Bill reacted to this like a horse who has suddenly seen something unpleasant and unexpected in the middle of the road. He removed his feet from the desk and said, 'Why? What reasons?'

'Only,' Parkhurst said, 'so that when I write that you've denied our application to stage the championship fight for the benefit of the *Mercury* Free Food Fund for Hungry Children, we can give our two million circulation cogent reasons for your refusal.'

Clay, his bald head gleaming in the shaft of afternoon sunlight penetrating into the State Building down-town, rasped, 'The *Mercury* always likes to get its facts right, Colonel. It's in your interests.' Usually Clay was as up tight as a nerve ganglion but now he was playing it Parkhurst's way.

'Well, reasons, reasons,' said Wildman, 'the Boxing Law—the regulations.'

Parkhurst said, 'Yes, I see. Do you know the Boxing Law in the State of New York?'

'Of course I do!' said Wildman. 'I administer it, don't I?'

'And very well, too, Bill,' admitted Parkhurst. 'I have a copy here with me,' and he produced it from his briefcase. 'It's a well-written law as far as any law can go regulating legalized manslaughter and it's served the State excellently for a long time. I won't take up your time with every paragraph about what it says you may do, but let's take it the other way around first and have a quick skim through and see whether there's any clause or paragraph which says that there could not be a championship fight of fifteen rounds between a man and a kangaroo.'

The Colonel bristled. 'Sure it don't,' he said. 'Why should it? What would be the crazy idea in writing such a law like that, that a man couldn't fight a kangaroo? Who would think of such a thing like that?'

'We would,' Parkhurst said, and the tight skin of Clay's face split into a rare grin. 'So you admit that there's nothing in the Boxing Law that prohibits you from sanctioning a bout between a man and a boxing kangaroo and, in particular, one who has already knocked out the champion of the world?'

'I—uh—no, it doesn't say that anywhere, but the fella that wrote that law—Senator Walker—didn't mean for people to fight animals.'

'Who, Jimmy?' Parkhurst said. 'He'd have loved it! If he'd been alive today, he'd be sitting in the front row, ringside, cheering for the kangaroo.'

The Colonel tried a different tack. He pleaded, 'Look, fellas, don't let's argue. You know that's how the law was written and intended and how it's been administered. I can't change it.'

Clay asked, 'Do you know when that law was passed, Colonel?'

'Sure! I—uh—uh—Well, it was quite a while ago.'

Clay said, 'It was in September, 1920, and do you know what year we're in now?'

'Why—uh—sure, I do—Anybody does: 1970.'

'So it's really the fiftieth anniversary of the passing of the Walker Law.'

Parkhurst now moved in, 'Do you know how many times in the last fifty years the Supreme Court has rendered decisions and interpretations of the Constitution and the Bill of Rights of the United States of America at variance with either the word or the intent of the Founding Fathers who wrote it?'

'I—uh—uh—no,' said the Colonel, feeling himself pushed into a corner again.

Parkhurst fished into his briefcase and took out four typewritten sheets on which some researcher had evidently

217

written up the decisions rendered. But as the Colonel quailed at the sight, he stacked them and put them back again, saying, 'Well then, we needn't go into that any more. Now let's look at the regulations covering the licensing of a boxer in the State of New York, or the qualifications necessary for participation in a fifteen round, championship fight.' Parkhurst consulted his copy of the law again and read, 'He shall be twenty-one years or over of age and of good standing in the community, with no record of having been convicted of any crime more serious than misdemeanor in any State, or have been sentenced to jail. He shall not have, or have had any connection with undesirable elements. He shall not have consorted with or known any convicted criminals or gamblers. He...'

The Colonel's fists suddenly crashed upon his desk top. 'Hey, wait a minute, fellas! You got the regulation right there! He shall be twenty-one years of age. So before a fighter is twenty-one, he's only allowed to go six rounds. I read in your column about how old your kangaroo is—eight years. I guess that's about it, ain't it fellas?' His voice went heavily sarcastic, 'You wouldn't want me to sanction a fight for an eight-year-old, would you—ha, ha!'

Parkhurst reached into his briefcase again and Wildman's little eyes followed the gesture anxiously. The look of triumph vanished from his face as the sports editor produced yet another typewritten sheet. He said, 'I have here a sworn statement from Professor Eldridge Jones, Curator of the Bronx Zoological Gardens, and former head of the College of Veterinary Surgeons, on the comparative ages of animals and humans. That is to say, the average longevity of members of the animal kingdom compared with the equal average three-score-years-and-ten ascribed to human life span, according to the Scriptures. You wouldn't go against the Scriptures, would you, Bill?' He consulted the sheet and said, 'For instance, take the domestic cat, *Felis Domestica*. According to the Eldridge Jones scale a cat who survived to the age of twelve or thirteen, would be the equivalent of a ninety-year-old person. One year of a dog's life,' Parkhurst continued, 'is equal to seven years of your life. Now, let me see, on the Eldridge Jones scale,' he fingered the list, 'H, I, J,

K ... Hmm. Kagu, kakapo, kalong, kamptozoa, ... Ah, here we are, kangaroo: Ratio of both the Australian gray and red kangaroos to man is four to one. Therefore Matilda's official and registered age would be thirty-two. Not too old for a fighter who has taken care of himself all his life, never touched a drop of liquor or smoked, and has kept in training throughout his career.' He handed the document to his companion saying, 'Here, Justus, you'll want this when you have the story written for the news section.'

The Colonel began to fuff and fluster again, 'Now, fellas, wait a minute!' he said. 'Before you write up anything...'

'Well?' said Parkhurst.

'Look, fellas, what are you trying to do, ruin me? What's going to happen to me if I give a licence for this bout after Philadelphia and Chicago Commissioners turned these kangaroo people down? What would the Governor say?'

Parkhurst replied, 'Philadelphia and Chicago are still living in the dark ages. This is 1970, Bill, remember? Where for twenty-five bucks anybody, including a minor, can buy a ticket for the theater to see two people humping on the stage. Something like ten million people saw Matilda knock out Cyclone Roberts on television, now they'd like to see what he does with Lee Dockerty. And, incidentally, contribute a kitty of two million dollars to the kiddies' Free Food Fund.'

Wild Bill was looking miserable and unconvinced. He said, 'I could get fired.' His job as Licensing Commissioner for Boxing was very dear to him, enabling him not only to indulge in his eccentricities, but to emerge from anonymity into the spotlight of the public press. Almost daily his name appeared in the paper. With his powers to say yea or nay to a proposed fight, he was legitimate news. Nor had he been too bad at his job and was noted for having refused to license mismatches that might have resulted in slaughters.

Parkhurst said mildly, 'Fire you? With the backing of the *Daily Mercury,* picture on page one of courageous Licensing Commissioner? Do you know what the *Mercury's* readership is?'

Wildman said, 'Uh ... Well, you said you had two million circulation...'

Parkhurst said, 'I'm not talking about circulation now.

Our ABC survey shows that there's an average of five readers to every copy of the *Mercury*—that's ten million people. Do you think the Governor is going to kick that many votes in the face when I print that we have been turned down for a charity bout for underprivileged children? You'll be the goat, Bill. The Governor will have to take the rap in the end.'

Wildman was still looking harassed but it was obvious that the ten million figure had managed to penetrate beneath his toupé.

'And anyway,' concluded Parkhurst, 'you've already licensed Matilda's manager and second.'

Wildman was really startled, 'Who? What? Who did? I ain't done anything of the kind! I wouldn't know them if I saw them.'

'Yes, you would,' Parkhurst grinned and referred to his briefcase again. 'On April 9th, last year, you licensed one Solomon Bimstein of 1620 Seventh Avenue, as manager of prize fighters, and William Baker of the same address to second prize fighters in the ring.'

Distant lightnings began to sear through Wild Bill's brain. He said, 'Solomon Bimstein works for Ahearn's stable. Patrick Aloysius came in to see me about it and I licensed him to work with...'

'Sol Bimstein, Patrick Aloysius Ahearn and Billy Baker are equal share owners in Matilda, with Bimstein as manager of record, so you might as well finish the job. Walter Mason is prepared to give us the Jericho Stadium for free. Matilda is willing to donate his full share of the purse. We'll be able to put some flesh on the bones of a hundred thousand hollow-eyed, undernourished kids and you'll be a hero.'

Temptation glittered before the avid eyes of Wildman. The way Parkhurst had put it there would be unlimited publicity. He might even be hailed as a pioneer. Parkhurst was right, 1970 was not 1920. Past Boxing and later Licensing Commissioners had taken stands, some of them unpopular, which stretched the rules. But a kangaroo? In Jericho Stadium? A sudden pang assailed him. He asked, 'What about the other papers, fellas? They could put me on the cross for going along with you on a thing like this.'

Parkhurst nodded and said, 'If the match was officially

sanctioned for our charity, they'd have to cover it. They'd be crazy not to. They might take a crack at you for allowing the bout, but once it was official, they couldn't ignore it, could they?'

Wildman made some rapid calculations and both Parkhurst and Clay were certain they could look right through the frontal wall of his skull and watch the brain waves pulsing within. Five papers to one, with a total circulation and readership greater than that of the *Mercury* all putting the zing on the Licensing Commissioner's office. It was a bad gamble. Wild Bill took his courage in his hands. He said, 'Sorry, fellas, I'm afraid not. Ask me anything else and I'll be glad to help you out. The answer is no. Maybe you could do what you done in Chicago—like actors on television...'

Parkhurst said evenly, 'We don't want to do what they did in Chicago. Matilda has fulfilled every condition to be entitled officially to the middleweight championship of the world under the laws and regulations of the State of New York if he can take Dockerty again.'

But Wild Bill had made up his mind. As the Boxing Law had been amended, the final decision was his and there was nothing anybody could do to him. The *Mercury* might write badly about him; it would be nothing new if they did. But on the other hand, four other major morning and afternoon papers would come to his defense. He said, 'Sorry, fellas, it's still no.'

He missed a glance flashed between Parkhurst and Clay and only to his relief saw the big sports editor get up in apparent defeat with, 'Okay, Bill, if that's the way you feel about it. You're the boss.'

But Clay did not arise. He said, 'Just before we leave, Mr Licensing Commissioner, do you know off hand how many times you've issued a licence to Lee Dockerty and his manager to take part in bouts in the State of New York?'

Wildman said, 'I dunno, maybe six or seven times. You can look it up. He was in two champeenship bouts...' He said, 'Why?'

Clay had made a note on the back of an envelope and said, 'Just for our story.'

In a sudden burst of irritation Wildman, who felt himself on firm ground, said, 'What's that got to do with your story? So you'll put the blast on me because I refuse to license a kangaroo? Maybe it's you fellas who'll look foolish.'

Clay said, 'Oh no, no, no, nothing to do with the kangaroo. Just the part about regulations.'

'What part? What regulations? What are you talking about?'

'The bit about an applicant for a licence as a boxer in the State of New York shall not have, or have had any connection with undesirable elements. "He shall not have consorted with any known or convicted criminals or gamblers."'

'What's that got to do with Lee Dockerty?' Wild Bill bristled. 'He's never been in any trouble and his manager Mr Schwab has been in the business for years, and never been refused a licence anywhere.'

'That's right,' Clay agreed, 'that's what our records say. But you ought to know, just as well as we do, that in actual fact Lee Dockerty's contract is owned by one Joe Marcanti for a third party known along Broadway only as Uncle Nono, Capo of Cosa Nostra's Eastern seaboard division. Did you know that?'

Wildman said, 'N-no.'

'Well, we do,' said Clay and arose. 'How about it, Bill?'

The collapse of Wild Bill Wildman was pitiful to see because in addition to the pouring sweat, the quivering mouth, the toupée went still further askew atop the skull inside of which, both the other men were now aware, was a chaotic merry-go-round of fear. He finally quavered, 'I don't dast, fellas. Go ahead and print anything you want about me. You can't make me say yes.' He tried to salvage some measure of dignity by getting to his feet but his knees were shaking so he had to hold to the desk as he said, 'That's all.'

On the way uptown Parkhurst said to Clay, 'Do we tee off on him? Do we print the story?'

Clay had relapsed into his sour mood again and he shook his head. He said, 'No. That's not what newspapers are for, to persecute people. You know, in a way, the silly bastard's got some guts to turn us down.'

Parkhurst asked, 'So what do we do now?'

'Drop it,' said Clay. 'End of the line. I said I'd go along with you if Wildman would give us a licence. Well, he hasn't. If Wildman's willing to take the rap, that's it.'

But when they reached the office there was a message for Mr Clay to call Commissioner William Wildman immediately.

Clay got on the phone while Parkhurst hung over him and listened to his editor say, 'Yes, Colonel. Yes, I see. You've thought it over? . . . It's for the kiddies, as you say . . . That's wonderful, Colonel . . . We'll announce it tomorrow. Many thanks.'

He hung up and said, 'Okay, Duke, you've got your licence.' He looked up at Parkhurst and asked, 'Now what do you suppose made Uncle Nono suddenly change his mind and give Wildman the green light?'

Parkhurst grinned and replied, 'Sheer curiosity, I expect. He's just as intrigued as we are to find out what will happen when Matilda gets Lee Dockerty back into the ring with him again.'

# XXI

BILLY BAKER BROKE the news personally to Matilda who being only a kangaroo, albeit the best 160-pounder in the world, said no more than, 'Uck-uck-uck-uck-uck,' and searched Baker's pockets for a sweet.

Patrick Aloysius had been out of the office when Parkhurst had telephoned Wildman's decision to grant the licence and that the fight was on, and Baker had proceeded alone down to the stables but stopping off to celebrate his forthcoming crossing of the threshold that divided million-aires from the non, at several oases known to him on the way, with a pint or two in each, so that by the time he reached Ryan's Stables and Matilda's suite he was feeling emotional.

'Well, you've made it now, you fuzzy barstid, you,' he said while Matilda munched ecstatically upon the Almond Bar, which Baker had picked up on the way so that Matilda might join in the celebration with him. 'You've got yourself a fight with the middleweight champ of the world and if you can zap Dockerty like you dotted 'im one before, we'll myke them lay a million dollars on the line if they so much as wants to shyke you by the 'and afterwards. Matilda from Australia, middleweight champ of the world, and Billy Baker of Bermondsey his bloomin' manager, that's the combination that's done it!'

Later he became sentimentally reminiscent, lying on his back with his hands clasped behind his head and thinking aloud to Matilda who was stretched out in the clean, fragrant straw, leaning on an elbow and regarding The Bermondsey Kid in the manner of one who is wholly bemused and fascinated with what he is hearing.

'You know, what I've missed all me life since I've been away from 'ome? The toots from the river at night. When I'm lyin' awake maybe, and can't sleep. The toots that used ter say, "Come on, Baker, get off your bum and go! There's the whole wide world to see." That's when I was a boy. Later them toots on the river of the boats and barges going up and down meant work on the docks, when I hadn't no fights and had to eat.'

He took a wisp of straw and chewed it. 'And there was the smell from the river—oil and tar and the river itself. The Thames is a bit high down about where I was born and on a hot night you knew it was there. I used ter like it in a fog, too. Cor, it weren't 'arf thick down there by the gas works, yellow and thick like, so the toots and bells from the ships could 'ardly get through. And there was always the smell of cabbage cookin' somewhere—good old British cabbage, there ain't nothin' like it nowhere else. You'd like a bit of that, Matilda.'

He reached over and pulled one of the animal's large ears cocked in his direction. The kangaroo gave an 'Uck-uck', leaned further on his elbow, nestled his muzzle into Baker's shoulder and gave a sigh of contentment.

'Washin' 'anging out on the line be'ind the 'ouses. Kids playin' and yellin' in the streets. The stink of coal smoke and the clatter of them big brewery drays rolling over the cobbles. That's what it was like then. Cor, what's it going to be like now?'

He paused to change the reels of his reverie. 'William James Baker, millionaire! Whoever would have believed it when I shipped out to your country, Matilda, thirty-five years ago. I went for two fights and I never come back. A great country, yours, Matilda, but it wasn't never home to me, now. 'Ome is where you was a kid and you never forget it.'

He contemplated the wisp of straw for a moment and said, 'I'll be able maybe to have it all now, what you can buy with money; house in the country, silk shirts, women, a Rolls Royce, if I wanted it. But I won't, you know, Matilda. You wouldn't see Billy Baker sittin' on 'is arse doin' nuffink. It ain't in me character. I've worked all me life and work's what keeps yer young. I'd buy back me grandad's old pub, The Bunch of Grapes, only you know what I'd do? I'd rename it. I'd call it The Punchin' Matilda. I'd 'ave the old board down and a new one painted showin' you with the gloves on and your dukes up. That would draw 'em in, especially with me workin' behind the bar—Billy Baker, the old Bermondsey Kid, undefeated lightweight champion of the British Isles, drawin' you a pint of bitter.

'They'd all come to see you too, Matilda. Gor blimey, if they wouldn't! I'd build you a bloody palace out back, made of hay and straw with bananas all up one wall and Hershey Bars all up the other, and a sign over the door, "Punchin' Matilda, Middleweight Champion of the World", where you could be taking your ease.'

At the words 'banana and 'Hershey Bar' Matilda raised his head and gave Baker an interested look, but when he saw there was nothing forthcoming, he collapsed again.

'Maybe,' continued Baker, 'in the mornin' before I opened, or at night after I'd said, "Time, Gentlemen, please", we'd put on the gloves together and have a bit of a go for auld lang syne's sake, until I got old and me legs got so stiff I couldn't get out of me own way, much less yours, you bleedin' punchin' bugger, what's going to make Billy Baker The Bermondsey Kid a bloody millionaire!'

The picture was too beautiful and in particular the thought of his growing too old to cope with Matilda brought tears to Baker's eyes. 'How would you like that, old boy?'

And then to his own surprise Baker turned likewise, leaned on his elbow so that he was facing Matilda and found himself saying, 'You wouldn't, you know. You wouldn't like it one bloody bit. That ain't no kind of life for a 'roo and the middleweight champion of the world. Billy Baker, you're a selfish bloody barstid thinking of shuttin' your pal up in a cage, just because you want to show yourself off in your grandad's pub. You know where he ought to be? Back in the scrub behind Brisbane where he come from, amongst the malee and mulga trees and the Hoop and Bunya pines, with 'is own herd, feedin' off the bush and hoppin' about where he pleased.

'Do you know what you'd be, Matilda?' Baker concluded,

warming up the new visions appearing upon the screen of his mind, 'King of the 'roos! Not just of your own herd, but king of every bloody 'roo in the whole of Australia. There wouldn't be one that could stand up to yer and you could 'ave your pick of any female that struck your fancy. Straight left; right cross, zap and Bob's your uncle! With what I taught you, there wouldn't be none that could come even close to you. They'd be talkin' about you from Cooktown to Brisbane, Brisbane to Sydney and Sydney to Melbourne. Matilda, the King of the 'roos who'll knock your bloody brains out! And articles in the *Sydney Mornin' 'Erald* about you, like in the old days, when we was barnstorming together.

'What was I saying before? Home is where you was a kid and you never forget it. Maybe you're longin' for the smell of eucalyptus and the bottle brushes, and to be back where you can start hoppin' and hop for six months without ever comin' to the end of space to hop. Australia's your home, Matilda, just like Bermondsey's mine and that's where you'd ought to be. It'll break me 'eart to see the last of you after all you done for me, but don't let nobody say The Bermondsey Kid can't be grateful. I'll take you back to the bush, Matilda, have a last look at Australia and maybe set up me billy-can once more, 'fore I turns you loose and goes back to London and Bermondsey. I'll 'ave your photograph up be'ind the bar and you'll never be forgot. Don't say no, Matilda, I've made up me mind. I'll take you back home before I goes to England. Look at you, lying there like you was really taking it all in. Matilda, King of all the 'roos in bloody Australia!'

Matilda loved the sound of Billy Baker's voice. He raised himself higher on his elbow, leaned over and gave The Bermondsey Kid a resounding kiss.


The staging of a World's Championship prize fight in such a huge stadium as the Jericho Arena, seating some hundred and ten thousand souls, entails an enormous responsibility and kept Parkhurst and his special staff in collaboration with Walter Mason, head of the corporation owning the gigantic Arena, busy from morning until night,

227

even though the fight scheduled for the evening of July 4th was two months away.

No doubt that the stadium would be filled, nevertheless there had to be constant promotion, publicity and ballyhoo. Tickets had to be printed, distributed and kept out of the hands of speculators, easily accessible training camps set up for Lee Dockerty and Matilda where the visiting press could enjoy themselves as well as cover the work-outs (Lee chose Atlantic City while Bimstein, Ahearn, Baker and Company selected a site close to Palisades Park in New Jersey, just the other side of the river from Upper Manhattan). A seating plan had to be laid out; special police and ushers rehearsed in their duties; box offices opened in various parts of the city and a press office established in The Grand Metropolitan Hotel in midtown New York, with a free bar adjoining for the convenience of funnelling information, statistics, news copy and alcohol into local as well as out-of-town journalists during the build-up for the big battle.

Applications for Working Press seats were received from as far off as Hong Kong and Tierra del Fuego which meant that the usual banks of four rows around the immediate ring had to be extended to ten. Australia alone was sending thirty reporters along with a block booking for five thousand seats. Matilda was not going to lack support from the home folks and the *Sydney Morning Herald* devoted a daily article to his cause.

Scaling the house to three million dollars was another problem which Parkhurst solved by inventing what was known as a Donor Section, namely the first five rows surrounding the Working Press, plush seats reserved for those who donated a thousand dollars or more to the Fund with special gold-embossed tickets and the name of the philanthropist filled in on each. Afterwards came ten rows of double A's, B's and C's, etc., at two hundred and fifty dollars. At a hundred dollars your 'ringside seat' marked Row A entitled you to twelve inches of pine bench twenty rows removed from where the action would be taking place.

The cheapest seats located in Outer Siberia on the farthest rim of the amphitheater were twenty-five dollars, except that Parkhurst was clever enough not to neglect Matilda's hippie supporters by setting aside ten thousand reserved benches at five dollars each for genuine specimens of revolutionary youth. San Francisco, Los Angeles, New Orleans, Pittsburgh, Chicago, Detroit and other major cities sent delegations of their bums, dropouts and weirdies to represent them; guitars and other musical instruments permissible.

The Society for the Prevention of Cruelty to Animals checked in and had to be satisfied. This Parkhurst accomplished by arranging for S.P.C.A. to have their own chief veterinary surgeon at the ringside with the power to stop the fight should it appear that Matilda was being subjected to any cruel and inhuman usage, a calculated risk but a necessary one to placate animal lovers.

There was also the matter of security for Matilda; special guards on twenty-four-hour duty at his training camp, even though Parkhurst had the feeling that Uncle Nono would not attempt anything until perhaps the very last minute and then in all probability the target would be shifted. Still the armed guards and Matilda's daily work-outs conducted in a bullet-proof glass enclosure surrounding the ring made for colour and copy.

Rival New York newspapers, after mounting a halfhearted campaign against Wild Bill Wildman for issuing a licence for the match, found themselves on the receiving end of scolding letters, succumbed to the general excitement, began assigning boxing and feature writers, and found the pay-off in circulation rise. There was hardly anyone who, was not interested in the eventual outcome of the battle, particularly when it became confirmed that Lee Dockerty was taking his training in deadly earnest, eschewing alcohol, depositing himself between the sheets by ten o'clock every night and doing his five miles of road work each morning to strengthen his legs.

Providing sparring partners for Matilda had also proved something of a problem, as he chewed them up, so to speak, at the rate of two or three a week. Having been out of serious action since the affair with Cyclone in Chicago, he was practically hysterical with joy at getting down to serious business again and could not be restrained from zapping the paid hands at least once during every workout. Sharp-eyed boxing experts reported his deadly accuracy and devastating punching power. At least one spar mate was hospitalized with a broken jaw and another had to retire from an injured kidney. The build-up and excitement mounted day by day as June drew to a close and the national holiday approached. Parkhurst made no mistakes. Self-confident and in full control of the situation he was riding high. The fight was a sell-out.

And Uncle Nono decided it was time to bring him down. The manner of it would in no way interfere with the forthcoming battle, it would merely make certain that Parkhurst would not be present and that newspapers rival to the *Mercury* would be guaranteed some sensational headlines.

The columnist had been correct in his estimation of why the shadowy, powerful Uncle Nono had apparently given his consent for the match to take place. It was the ex-college sportsman and athlete who had been unable to resist. At the same time, the executor of Cosa Nostra's policies could not afford to neglect the punishment of those responsible for putting him in an invidious position.

The morning of the 1st of July, Gio di Angeletti summoned Johnny Renato, his right-hand man, who entered his office, his hands filled with the glitter of gold. He put down ten Donor tickets on Gio's desk and said, 'Here you are, boss, front row. They cost you ten grand.'

Di Angeletti shuffled through them and said, 'Fine! Two of these are for you and the missus.'

'Thanks, boss. Any final instructions?'

'Yes. Joe Marcanti; he hasn't been exactly a firecracker, has he? And I can't see him of any more use in the future. He will make a useful example. Hit him.'

'Yes, boss.'

'With regard to Parkhurst, have you got all the dope?'

'Yes, boss.'

'Okay, lower the boom.'

And finally Gio added, 'Those three—Bimstein, Ahearn and Baker—I want you to see them.'

'Like where, boss? You mean here, or—the other office?'

'Perhaps downtown,' said Gio di Angeletti, 'although when we're through with them, they won't be talking.'

The timing, Parkhurst had to admit, had been superb. Ever since the licence for the fifteen-round championship fight for the benefit of the Free Food Fund had been issued he had been wondering not so much how, but when Uncle Nono would strike, since strike he must if only for the sake of his prestige in his own organization, not to mention personal pique. Lately in the confusions of all the details to be worked out he had quite forgotten.

The two F.B.I. agents presented themselves in Parkhurst's office as he was wrestling with last-minute demands of heavy advertisers for seats in prime locations. It was four o'clock of July 3rd, too late for the afternoon papers but ideal for the mornings of July 4th when news would be scarce.

The two young and serious-faced gentlemen carrying briefcases identified themselves as agents of the Federal Bureau of Investigation and in spite of himself Parkhurst could not repress an inward shudder of apprehension. Had he slipped up somewhere? The very innocence of the conservativeness of their clothes was more ominous than if they had been blue-uniformed and armed.

The first F.B.I. man said, 'I'm afraid, Mr. Parkhurst, that we are here on a most unpleasant errand and one we would greatly prefer not to be involved in. May I ask whether you are familiar with a Federal Law known as the Mann Act?'

Parkhurst said, 'I've heard of it in a general way.'

The second F.B.I. man said, 'Mr. Parkhurst, before we arrest you for violation of the Act, we are prepared to go over some of the facts with you. You need not comment upon them if you don't wish to.'

Parkhurst looked upon the pile of tickets, orders and memos on his desk and said, 'That's all right. Go ahead.'

The first F.B.I. man went into his briefcase and produced a whole file of papers containing what looked like photostats, carbon copies and affidavits of one kind or another. He leafed through them to see that they were in order and began, 'Mr Parkhurst, on September 8th, 1969, you purchased two, first-class, round trip tickets to Philadelphia and two pullman seats. You were accompanied by a Miss Birdie McBride of 315 West 48th Street, the Claymore Apartments. I have here an affidavit of Morton Wile, ticket clerk of the Pennsylvania railroad, who remembers selling you the tickets on that date. You checked into the Quaker State Hotel, a double room, which later was occupied by you and Miss McBride. We have witnesses to the effect that she spent the night in the room with you.'

Parkhurst said nothing and the agent continued: 'I will now show you a copy of the record of United Airlines which indicates that on November 5th, you purchased and paid for two tickets to Chicago on flight 407. These carbon copies of the tickets prove that the flight was booked in the names of Miss Birdie McBride and yourself. In Chicago you registered into the Ambassador West Hotel with Miss McBride, occupying Suite 710-12, with a sitting-room between, where you remained for three days. I have here an affidavit of Ralph Wasinski, a room service waiter, who deposes that on two mornings—November 6th and 7th, he served breakfast to you and Miss McBride and that you were occupying the same bed together.'

Parkhurst could not refrain from murmuring, 'How continental.'

The F.B.I. man produced another paper. 'Here is a deposition from Madge Somino, chambermaid in charge of your suite, who declares that at no time had the bed in Miss McBride's room been slept in.'

Parkhurst said, 'Naturally,' and both men looked at him sharply, and one said, 'You see, unfortunately, Mr Parkhurst, your face is easily recognizable and you are a very well-known person.'

The second agent asked, 'Do you wish to deny any of this?'

Parkhurst replied, 'No.'

The agent went on, 'You and Miss McBride continued to occupy these premises until the morning of November 8th, when you flew back together to New York on United

Airlines flight 962, returning with her to your apartment. May we assume this to be correct?'

'So far,' Parkhurst said and then added, 'Since when have you chaps taken to working for the Mafia?'

Both men stiffened perceptibly and one said angrily, 'Mr Parkhurst, the Federal Bureau of Investigation does not work for anyone but the taxpayers who support it and the laws that it protects.'

Parkhurst said, 'Come off it, boys! You don't go around collecting carbon copies of airline tickets from private citizens until there's been a complaint, or there's word of a big white slave operation going. There must be a couple of hundred thousand girls a night transported over one State line or another for a roll in the hay in some motel, but until somebody fingers the guys who transported them and there's a bit of blackmail involved, you never hear about it.'

The first F.B.I. man said swiftly, 'Mr Parkhurst, the F.B.I. doesn't negotiate with criminals.'

Parkhurst said, 'Since when? What the hell would you boys do if you didn't get a canary to chirp every so often? Whoever gave you the tip probably supplied you with some of the material as well, or told you where to go to check up. It's all true. The only trouble is that you didn't start soon enough.'

He opened the center drawer of his desk and took out an envelope which he held between his fingers. He said, 'On the morning of the 17th August, 1969, Miss McBride and I crossed three—count 'em, three—State lines together: New Jersey, Philadelphia and Maryland, to Elkton. I paid the railroad fare. At Elkton we took out a marriage licence,' Parkhurst jiggled the envelope, 'herein contained, issued by one S. M. Bootley. We then visited a Justice of the Peace, C. J. Marshall, were married and were given a marriage certificate, also herein contained,' and he shook the envelope again. 'We thereafter returned to New York. Hence I should say that from then on and during all subsequent trips we made, we were man and wife.'

The second F.B.I. man lost his head. He said, 'You admit you crossed three State lines!'

Parkhurst asked, 'Since when is getting married an

immoral purpose? Anyway, there wasn't time. We had to be back by seven o'clock. Miss McBride had a performance that night. Actually, by the time she had finished, we were both too exhausted for immorality. It had to wait until the following morning.' And then he snarled, 'What the hell is the matter with you fellows? You check out everything on a supposition of guilt and don't even allow that there might be a case for legitimacy. You're just lucky that I didn't let you make an arrest. You'd have been a laughing stock and it would have cost you a suit besides.'

The first agent said, 'Why did you continue to use her maiden name? Why didn't you register as Mr and Mrs Parkhurst?

Parkhurst said, 'Birdie McBride is my wife's professional stage name; it has a value for her. Besides which, it didn't suit our convenience to have our marriage revealed at that time. When I'm ready to have it announced, she will be known as Mrs Duke Parkhurst,' and he added with a sudden grin, 'One Lucy Stoner was enough.'

The first agent pointed to the envelope in Parkhurst's fingers and said, 'May we have a look at that material?'

'No, you can not!' said Parkhurst. 'You birds love to travel. Travel some more and dig it up for yourselves. Since, as you say, my face leaves indelible impressions, the ticket clerk at Penn Station will remember me, and so will the J.P. and Licence Bureau clerk in Elkton. They ought to, after the tip I gave them to keep it secret.' He opened the desk drawer and dropped the envelopes inside and shut it again.

The first field man returned the papers to his briefcase, snapped it shut and the two arose. One said, 'Our apologies, Mr Parkhurst, and may I say how pleased we both are that the charges that were to be laid against you are without foundation, and ...'

'Oh, Christ!' Parkhurst said. 'What a world we live in. You'd have been just as happy if the case had been as open and shut as you thought it was, and you had laid Duke Parkhurst by the heels. Big feather in a small cap. You and your informant were already reading the headlines, weren't you? "DUKE PARKHURST, MERCURY EDITOR, CHARGED WITH MANN ACT VIOLATION. ACCUSED OF TRANSPORTING BROAD-

WAY DANCER ACROSS STATE LINE FOR IMMORAL PURPOSES. F.B.I. MAKES ARREST IN MERCURY OFFICES. CHAMPIONSHIP CHARITY FIGHT TO PROCEED AS SCHEDULED."'

The two agents said no more and left. Parkhurst wiped some perspiration from his forehead, for the headlines he had conjured up might indeed have become living ones as he reflected upon the things men did against all reason and intelligence in pursuit of their momentary passions, and he said to himself, 'Uncle Nono just didn't delve far enough into the history of my first marriage, or guess that I was bound to fall in love for keeps with someone as dear, stupid, inarticulate and irresistible as Birdie McBride.'

# XXII

THE NIGHT OF Saturday, July 4th, 1970, was clear, starlit, with a moon ranging across a milky sky. The day had been hot; the night was breathless but bearable. A hundred and ten thousand souls jammed into the giant bowl of the Jericho Stadium sat crouched in the dark beyond the lights shed by the canopy of arcs over the ring.

A great fight crowd creates and animates something tangible over and beyond its own menacing presence, something disturbing and frightening. Its members have one thing in common: they have gathered there in the hopes of seeing someone injured, if possible severely, before their eyes. This mass lust for violence centers upon the tiny square of light at the bottom of the vast circular amphitheater and engenders an atmosphere of threat and evil that is almost palpable. It attacks the nerves in one's stomach and one feels one's heart beating high in one's throat. It arouses unreasonable anxieties and heavy foreboding.

The preliminary bouts were over. The appetizers served up in the pairing of a half dozen of the younger middleweights had been satisfactory in producing strife, blood and a knock-out. The crowd was warmed up. Now it was awaiting the main event to come into the ring. The pool of light at the center was a-buzz.

Here in the Donor, pride-of-wealth charity Section, sat the great and the would-be greats; millionaires, stage and screen stars, film producers, politicians, gangsters, owners of Broadway eating houses, newspaper tycoons, society sportsmen—a cross section of the riches and importance of metropolitan New York.

236

Four steel platform towers raised on each side of the ring, far enough back so as not to interfere with the view from the rearmost seats, were black with television, moving picture and still camermen. Other photographers with hand equipment along with reporters and feature writers invaded the Donor Section photographing and interviewing the celebrities. Thousands of moths and insects fluttered beneath the canopy of lights. The Working Press in shirt-sleeves were sweating and chatting nervously. White-coated candy butchers infiltrated, stridently yelling their sales cries of 'Get your ice-cold drinks here!'

That year's official gate-crasher, a filthy tramp by the name of Dog-Face Digby, paraded himself at the ring-side. Parkhurst had given instructions that at whatever gate he presented himself with his usual pail of water, he should be admitted. He lent colour if not odor to the scene with his cap askew over one ear, torn sweater and patched pants.

In the aisle between the last Working Press benches and the first row of the Donor Section the celebrities visited or shouted to one another their 'Hi's!' and 'Hello, there's', and 'Who do you like?', to make sure that their presence was being noted in the list of those of importance.

There was the firefly glow of matches lighting expensive cigars. Exotically gowned women, socialites, actresses, heterae, their cunning cat's-eyes quick to catch the gleam of the photographer's lens aimed at them, turned on the smile and the best side of their faces. From the darkness of Outer Siberia, the section where the hippies were gathered, came strains of guitar music and young voices lifted in song, mostly obscene.

As promoter of the championship, chairman of the charity, editor and columnist, Duke Parkhurst in dinner-jacket roamed the area checking, greeting friends, savoring the success of this venture that had started as a gag. He passed Giovanni di Angeletti sitting in the front row with his stunning, actress wife Leileen, his family and party of friends.

Di Angeletti said, 'Hi, Duke!'

Duke Parkhurst said, 'Good evening, Mr Angeletti.' They knew one another. Di Angeletti was in the front row at every important sporting event.

The distinguished importer was wearing a moiré dinner-jacket and a blue, ruffled shirt. He arose to speak with Parkhurst who towered over him. 'It's a great job you're doing here, Duke. Just great for those kids. I've sent in my cheque for ten thousand for our seats.'

Parkhurst said, 'Thank you, Mr Angelletti. Yes, I know, you've always been very generous with us. It's people with hearts like yours who are the sterling citizens of our city.'

The fire at the end of di Angeletti's stogie glowed brightly for an instant and that was all. Then he asked the question that was still going the rounds whenever two men met and exchanged greetings, 'Who do you like?'

'I've picked Matilda in two,' Parkhurst said.

Di Angeletti's cigar end glowed again, 'Could be, could be,' he said and sent his own mouthful of blue smoke drifting up to join the haze settling around the area. 'If he doesn't take him in two, Lee might have a chance. I hear he's trained for this one.' And then he added, 'You could be right. I saw a re-run of the Chicago fight on television the other night.'

Parkhurst knew of the rumors that in some way had been trying to connect di Angeletti with the shadow empire of Cosa Nostra, but was one of those who thought it ridiculous. Although younger than di Angeletti, he remembered reading of his exploits on the Harvard football squad and was a long time aware of him as one of the well-known group of wealthy New Yorkers, first-nighters and sportsmen. It just did not figure. And yet Uncle Nono existed. There must have been at least a hundred people at the ringside who thought they suspected who he might be and a dozen or so, including the Police Commissioner, the District Attorney and the Regional Director of the F.B.I., who were actively engaged in investigation.

In the news columns of the papers of that very morning was the mystery of the passing of yet another prominent Mafioso whose body was found down an abandoned mine shaft in Montana, the enigma being compounded by the fact that the head that fitted it—that important component being missing from the corpse—had turned up in a Pan American plastic overnight flight bag in a luggage locker at the railroad station of Flagstaff, Arizona. Not yet fully identified, the two parts were thought to belong to one Joe Marcanti, a small-time New York hoodlum involved with boxing.

If this turned out to be true, Parkhurst thought, the fact coupled with the visit he had had from the F.B.I. was an indication of Uncle Nono's growing irritation. What might he be capable of if Matilda in a few minutes publicly flattened his boy, Lee Dockerty.

Looking down upon the elegant, self-possessed patron of the arts, sports and charities, Parkhurst thought that if he really was Uncle Nono he had an impudence that verged upon the magnificent. He asked, 'Are you betting?'

Di Angeletti at his most charming, smiled and said, 'I don't bet on fights. Horses are my weakness. Well, congratulations again,' and he sat down.

As usual, every kind of rumor about the fight had been voiced and passed about the ringside in the final moments as the insiders, the wiseguys, the tin-horn gamblers and Broadway sports who wanted to be thought in the know, gave them last-minute, side-of-the-mouth circulation: Uncle Nono had bagged the referee. Patrick Aloysius Ahearn had bought the judges. Lee Dockerty was going to take a dive. Somebody had managed to slip some dope into Matilda's breakfast food. Lee Dockerty had hurt his left hand in training but the news had been kept from the public. If the going got rough, Dockerty was going to foul out. Both sides had bribed the timekeeper to lengthen the rounds/shorten the rounds/give whichever fighter hit the deck a long count/a short count/no count. Matilda was not a kangaroo at all but a very clever boxer dressed in a kangaroo skin. But the odds you could get, seven-to-five either way and three-to-one on a knock-out, did not reflect the truth of any of these.

A handsome, square-jawed man with wide-set eyes and short-cropped grizzled hair, wearing gray trousers and gray shirt, the prescribed uniform for New York referees, ducked through the ropes and Parkhurst eyed him with satisfaction. Colonel Wildman might be a fool in many respects, but he was selecting his officials for the night with particular care. The referee was Honest Phil Gahagan, for fifteen years an arbiter of impeccable integrity with never so much as a blemish on his record. The officiating would be clean.

The presence of the referee in the ring presaged the fact that the main bout was ready to come in. Parkhurst climbed over the rows of press seats to his own which was so close to the action that when he squeezed his bulk into the narrow place, his eyes were just at the level of the canvas-covered ring floor. Matilda's corner was on his right; Lee Dockerty's opposite on his left. O'Farrell sat at his side to send the round-by-round.

Quiet fell upon the nearest spectators and there was no sound but the chattering of the telegraph instruments as reporters dictated their running stories for their early editions. The drama was about to begin.

A rolling cheer erupted from the far end of the bowl and followed the entrance of Lee Dockerty accompanied by Pinky Schwab and his handlers, as they hurried down the aisle. When they climbed into the ring and took possession of their corner, it waxed into the full-throated roar of fans welcoming the champion.

Through one of those curious switches in public affections Lee Dockerty, who had never aroused a great deal of emotion amongst his followers, had suddenly become popular. From the Donor seats came the rattle of applause, but those out of sight in the darkness stood up and cheered him.

Lee Dockerty, got up off his stool, danced about and clasped his hands over his head. He was clad in white trunks. Over his shoulders was a blue silk bathrobe with 'LEE DOCKERTY MIDDLEWEIGHT CHAMPION OF THE WORLD' lettered in gold on the back. His manager and two handlers wore blue sweaters with the same legend. The welcome was so deafening that Parkhurst had to scream his dictation into the ear of his telegraphist.

And now on the other side of the stadium there arose a low murmur, an 'Oooooooooh!' sound that combined astonishment and nervous laughter that rippled down to the centre of the arena in the wake of the arrival of Matilda with his entourage of Bimmie, Patrick Aloysius Ahearn and Billy Baker, the Bermondsey Kid. All three were clad in white sweaters on the back of which was a small replica of the flag of the Australian Commonwealth with its Southern Cross

240

and larger star under the Union Jack in the upper left-hand corner on a field of blue and in red letters, 'MATILDA MIDDLE-WEIGHT CHAMPION OF THE WORLD.'

There were no cheers, only the distinctive rushing sounds of thousands of people on either side of the aisle standing up to get a better look, for Matilda in a red, white and blue, brass-studded collar with The Bermondsey Kid holding his chain, was hopping along low to the ground. After a moment, from somewhere in the gloom beyond the centre lights, an Australian contingent gave vent to a 'Coo-ee! Coo-ee!' and from the hippie section came a burst of music and one great, obscene yell at Dockerty. But the sound failed to fire the crowd and it died away.

Then came the shock, a half shout of disbelief and again the nervous laughter, as with his characteristic leap over the ropes, Matilda bounded into the enclosure and suddenly erect, stood in his corner.

Although the crowd had been expecting it, since all had heard of him and more than half of them had seen him in action on television, it nevertheless came as a shock. In one corner a man—a superbly muscled athlete, blue-eyed, dark-haired, jigging to loosen up—and in the other, a beast and a grotesque one with a barrel chest, powerful shoulders, short, thin arms, a mighty tree-branch of a tail and the head of an ass. It was unreal, impossible, unacceptable and yet for that moment when Matilda stood erect and turned his glance from left to right as though counting the house, he was a figure of disquieting menace.

Duke Parkhurst, over whom he loomed like something gigantic from prehistoric times, had a sudden qualm, a fervent wish that he had never started this, coupled with a flash of insight as to why the spectators had made the ground shake with their thunder of cheers for Dockerty and, except for the lunatic hippies and the Australians, had given no welcome to Matilda. They were frightened. A creature of a lower order was challenging the supremacy of a human in ordered combat. What if kangaroos were suddenly to become the master race and take over the earth? The reaction of the crowd, its hostility expressed by the strange nervous giggling, was as definitely racist as though Dockerty were preparing to fight a black man before a wholly white audience except that Matilda with a curiously supercilious look on his long donkey face, the ceaseless twitching of his huge, rabbit ears and the weird proportions of his body, inspired fear. Women in the audience clutched their escorts. There were many there who wished they had not come.

There was disbelief, too, in the utter strangeness of the scene, one which had never happened in any prize ring before or in any combat arena since Roman times, or even Regency days, when at fairs men wrestled with bears. Yet there it was and The Bermondsey Kid was winding soft bandages around Matilda's paws before donning the six ounce gloves, beneath the scrutiny of Pinky Schwab who had come over from Dockerty's side to watch over the process. Patrick Aloysius was performing the same function in Dockerty's corner.

There were those in the audience who had to remind themselves that stranger things than what they were witnessing had characterized the beginning of the latter half of the twentieth century: men had walked upon the moon; life had been gestated in test tubes; science had shattered every old-fashioned human concept and the century had produced—well, then, a kangaroo which had turned out to be the greatest and most scientific boxer of all times for his weight and size, one who had knocked out one world champion and two ex-title holders. And some in the crowd turned to more homely consolations. Even before this day and age, cowboys had bull-dogged steers in a test of skill and strength between man and beast; matadors in Spain fought wild and vicious bulls; in Florida swimmers entered tanks and wrestled with alligators; in circuses trainers grappled with lions and tigers so here was a kangaroo about to lace on red leather, six-ounce boxing gloves...

Pinky Schwab leaned down closer to make certain that nothing extraneous was being slipped into the gloves except Matilda's bandaged paws. The kangaroo reached up lovingly, put both arms around his neck and gave his face a thorough licking to the stunned silence of those down front who could see and through which came Baker's voice, 'Here, Matilda, behave yerself! I'm sorry, Mr Schwab, but just before he zaps 'em 'e loves everybody.' A shout of laughter dissolved some of the tension.

242

Only for Parkhurst it increased the strain and he was unable to keep his thoughts from turning back to those he had entertained a few days before when, alone with the beast, he had talked to it and tried to solve its mystery. He realized that he was undergoing the same pangs of nervous apprehension that assailed him at every important fight he covered, as though Matilda were a human being. He had become attached to the big, furry animal and it was always like that when writing boxing, impossible to be neutral. One formed a liking for one or the other of the contestants; one associated with them in training camps chatting or playing cards after the work-out. One knew about their lives, their problems, their wives and when one attended them at their moment of trial, it was something like going to see a friend have an accident. Death was not an unknown visitor to the arena, or to a hospital bed afterwards. Matilda was probably the most perfect, natural fighting machine he had ever seen, but that roped-off, canvas covered platform just above his eye level was also one of the loneliest places in the world once the bell rang.

And that bell was clanging now, to still the murmuring of the crowd for the announcer who had crawled through the ropes, carrying his microphone attached to a long cord. He waved his arms for quiet.

A half-dozen fighters, their suits tight around their bulging muscles, jumped into the ring to receive fulsome introductions or challenge the winner. Each one had to go through the ceremony of visiting each corner and shaking hands with the contestants. Two of them got affectionate licks from Matilda.

Parkhurst felt his nerves about to break and he muttered audibly, 'Oh, for Chrissakes, get on with it!' The load of responsibility, for once, was almost too much for even his huge frame to bear.

The last introduction completed, the bell jangled once more, the crowd hushed and Parkhurst, via the ganglia at the back of his neck, felt how much more menacing were over a hundred thousand souls in dead silence.

'Ladeez and gent'mun! In this corner, from Camauga, Mississippi, the middleweight champion of the world weighing 159 pounds—Lee Dockerty!'

Lee was up and dancing about, waving his gloved hands to the shattering roar of acclaim that followed his introduction.

Parkhurst turned to O'Farrell and said, 'Thĕy want Dockerty to win. They're for Dockerty. What will they do when Matilda flattens him?'

The little boxing writer was having his own collywobbles. He said, 'What will *we* do if he doesn't?'

The bell stilled the cheering. The announcer could be heard again, 'In the opposite corner, from Sydney, Australia, at 160 pounds, his worthy challenger—Matilda!'

There was a doubtful patter of applause from the dinner-jacket and evening-gown section and again that long 'O'oooooooh! and nervous laughter. What did one do, what kind of noise did one make to indicate to a kangaroo that was for him?

A burst of twanging music and the 'Kiss of Death' song, to which some more ribald verses had been added, came from the hippie section to be submerged by violent booing from the banked-up cheap seats in the far darkness which drowned out the cheers of the Australian contingent.

Parkhurst said to O'Farrell, 'There's your fight crowd for you. The last time out the bastards booed Lee.'

O'Farrell asked, 'Is it going to worry—our boy?'

'I wouldn't know,' Parkhurst replied. 'How can you tell what does or doesn't worry a kangaroo. Baker says he always is keyed up when he fights in new surroundings. Look at him.'

Billy Baker and Patrick Aloysius had helped Matilda lift off his stoop to acknowledge the introduction. As he had been taught, he clapped his leather-encased paws over his head and then stretched his arms along the angle of the ropes. One ear was pitched forward, the other angled backwards giving the effect from a distance as though he was wearing a huge bow. Bimmie and The Bermondsey Kid were stroking his back gently.

For a moment, Matilda glanced down at where Parkhurst and O'Farrell were sitting and the columnist for an instant looked directly into those brown, trusting eyes. He felt that not only did he see a gleam of interested recognition come into them, but that in some manner he was able to read the animal's thoughts and so excited was he that he heard himself call up to him, 'Forget your goddam Hershey Bars, and keep your mind on business!'

O'Farrell said, 'He ain't human.'

Parkhurst snapped testily, 'Of course he isn't, you ass! But he isn't worried either. Look at him.'

From somewhere in the rows of the Working Press a voice said, 'Go on out there, you lop-eared son-of-a-bitch, and get what's coming to you!' For the first time Parkhurst was aware of the hostility that surrounded him. All his colleagues, those from out-of-town as well as the local, except for a few professional friends, were wishing him ill. He had brought off the circulation, publicity, charity coup of the year for his paper. He had got away with the sanctioning of a boxing match between a man and a beast and, what was more, he had compelled them all to accept it and cover it. Most of them, and their editors and publishers as well, were hoping for his downfall, some catastrophe that would enable them to take a high-minded, editorial stand and bring him, his pride and the position of the *Daily Mercury* into the dust.

He thought to himself. *My God, what if Matilda knocks him out with the first punch? These people have paid three million dollars to see a fight and it could all be over in fifteen seconds!* Then he tried to soothe himself with another thought: *They've done nothing of the kind! They've paid their three million dollars in order to be here, so that when whatever is going to happen, happens they can say they were there and saw it.*

The announcer left. The referee motioned for the two contestants to come to the centre of the ring.

Lee Dockerty walked forward with his graceful, tiger's gait, pushing one glove against the other to work the padding away from the knuckles. Matilda made it in one hop, bringing forth another rustling and murmuring.

Dockerty was accompanied by Pinky Schwab and a handler. Matilda was chaperoned by The Bermondsey Kid and Ahearn. Bimmie remained outside the corner, leaning on the ropes. He was pale. Parkhurst had the impression that something other than the fight was worrying him.

The columnist called up the most futile questions to him, 'How is he? Is he all right?' But Bimmie did not reply. He did not seem to be concentrating on Matilda. Instead his glance was wandering around the Donor Section as though searching for someone and when from his vantage point he had apparently found who he was looking for, he licked his lips and swallowed. Parkhurst could not see who it was but not knowing of the emotional disaster that had befallen Bimmie, thought it was probably his girl and her family, the one whose mother was a Woolf.

A microphone had been lowered over the centre of the ring so that snatches of the referee's instructions to the contestants could be picked up and broadcast over the arena.

'... Lee, you know the rules. I want you to break clean; when I say "Break", step back. In case of a knock down, go immediately to a neutral corner and stay there until I signal you to resume. You know that in accordance with the laws of the State of New York, if you are knocked down, whether or not you get up immediately, there will be a mandatory count of eight. Keep your punches up above the foul line ...' Here he eyed the usual rope around Matilda's middle for a moment and suppressed a smile. 'I want a nice, clean fight.'

He turned to Matilda and said to Baker, 'Now what about this fellow? Does he know that ...?'

'He knows the rules,' Baker concluded for him. 'He's been trained right and proper, Marquess of Queensberry. All you need to do is to tap 'im on the shoulder and say, "Break". 'He's got ring manners. You won't have any trouble with him.'

The referee nodded and said, 'Okay, then, shake hands ... er ... boys, now, and at the bell come out fighting!'

# XXIII

DOCKERTY AND MATILDA touched gloves. The latter also reached over and gave his opponent one loving kick.

Dockerty said, 'You stinking bastard!' and returned to his corner, wiping his face on his arm. There was laughter and then the crowd settled down into its last moment of anticipatory hush.

With his back turned to his opponent, his seconds whispering last minute instructions, Dockerty shuffled his feet in the rosin. At the bell, he turned swiftly in fighting pose, his left out, his jaw covered, to move towards the center of the ring. But Matilda was already there, having made it in one leap. He stunned Dockerty with a left hook to the ear and followed that with a short right chop to the jaw and Dockerty went down on the seat of his pants with a thud, one leg curled up under him. And then Matilda was in a neutral corner, his arms stretched along the ropes, nostrils still twitching slightly. If anything, the expression in his eyes and along his muzzle was one of disappointment, as though somehow he was being robbed of an evening's entertainment.

Against the yell that broke forth, the sound of more than a hundred thousand people leaping to their feet to see better, the hammer of the knock-down timekeeper beating out the count could only be heard in the front row of the Working Press where Duke Parkhurst, also on his feet, was groaning, 'Oh, my God! The second punch!' And then, remembering, he shouted to his telegraphist, "Dockerty's down from a one-two to the head. He doesn't know where he is . . . Matilda caught him cold. I think it may be all over...'

The referee had already seen that Matilda was lounging nonchalantly in a neutral corner, minding his own business. He went over to the fallen boxer and picked up the count from the timekeeper, also indicating with fingers before the glazed eyes of Dockerty the number of seconds, '...Four...Five...Six...'

At seven, Dockerty had managed to get his leg out from under him and twist around on his hands and knees. At eight he shook his head hard and at nine, he was on his feet. Gahagan gave him the customary few more seconds of respite by wiping the rosin dust from his gloves onto the front of his own shirt and then stepped back and motioned Matilda to come out of the corner.

Dockerty fell into a clinch and got his face licked. He clung on, raging and cursing. But Matilda did not attempt to punish him. He wanted some fun. As it had with several other opponents on other occasions, the big rasping tongue of the kangaroo and the moisture had the same effect of reviving Dockerty as though he had been in his corner with his seconds dashing water in his face with a sponge. When the referee managed to pry him loose and he stepped back, his head was clear again and all his wits were functioning. As wave after wave of sound poured over him, he did the only sensible thing there was to do. He ran.

He did not even attempt to strike a blow but back-pedalled, ducked, covered up, used the ropes for added speed to swing himself out of range and from the mass roar, isolated cries of encouragement reached through to him: 'That's it, Lee boy! Keep away from him!' 'Don't let him catch you!' 'Keep moving, Lee!'

And move he did, with all of a veteran's ring guile and craft, speed and instinct. He had trained for this fight. The trembling was gone out of his legs. He ran and ran.

Parkhurst, thanking his stars that the fight had not ended in the first fifteen seconds, tried to settle down and get his heart back down out of his throat. Against his will he found himself hoping that Dockerty would make it. The pursuer and the pursued. Sympathy automatically chose the latter.

Matilda was puzzled. Parkhurst could see this, or thought

he could and, he imagined, disappointed as well. There was a noticeable wrinkle in the fur of his brow, just above the eyes as he contemplated the entirely novel problem of an opponent who was making no attempt to attack him but just ran backwards. He tried to pursue Dockerty diligently and Parkhurst could see that Matilda was studying the situation. Once he caught Dockerty and whaled him in the side and later again hit him with his whip-lash left hook, but Lee was going away at the time and not hurt. Such was the uproar that the timekeeper had to keep clanging his bell to indicate the end of the round. Lee, by fighter's instinct, was close to his corner where he collapsed onto his stool.

The cheering had subsided once more to the buzzing of excitement again. In Dockerty's corner Schwab was babbling instructions into his ear, while his seconds massaged his legs, lifted his rib cage to help his breathing, ministered to him, and it was evident to those close up that Dockerty's eyes were clear again and he was back in full possession of all his faculties.

In the opposite corner Matilda had refused to sit upon his stool and stood with his arms along the ropes, shaking his head, while looking over at Dockerty with what Parkhurst thought was almost indignation at such unsportsmanlike tactics. The Bermondsey Kid was stroking his back and pouring advice into one ear, 'Easy now, boy. Wait for him. He won't get away next time. Make him come to you. You've got plenty of time. There ain't no place he can hide, he can't run all night.'

A sardonic voice from the Working Press section called up, 'Attaboy, Billy! You tell him!' Parkhurst saw that Patrick Aloysius and Bimmie were having a whispered conversation.

The second round was a replica of the first, except that Dockerty did not hit the deck. Evidently he wanted to recover himself further from the numbing effects of being dropped before he was properly warmed up. His handlers at that moment being interested mainly in his survival, had told him to continue to run. In view of the first round knockdown, the fans would be patient with him if he refrained from exposing himself to another such blow and as he slithered about the ring, side-stepping, ducking, keeping tantalizingly just out of reach, he thought that he had discovered a weakness in his opponent that might be exploited later. The furrow in Matilda's brow took on an added ripple. Parkhurst suspected that in addition to the tactical problem to be solved, he was not having any fun. It took two to make a fight; the other kangaroo was just not playing the game.

They were onto it in Matilda's corner as well. Bimmie asked, 'What's the matter with him? Why don't he go in and knock his brains out?'

Ahearn said, 'Boy, this one is going to stink the joint out!'

The Bermondsey Kid said, 'He'll be all right. He's got to come to him sometime.'

The crowd was giving tongue again. Murray O'Farrell between dictating the round-by-round to his operator, '... Dockerty is still running. Matilda is trying to corner him but can't catch him...' yelled up at Dockerty. 'Stand still, you tin-canning son-of-a-bitch!'

But Parkhurst had analysed Matilda's problem just as Dockerty had. In a stand-up boxing exchange, balanced on his powerful tail, Matilda was agile as an eel and as hard to pin down. Instinct, training and practice coupled with his own particular genius for the sport allowed him to react within a fraction of an inch or the split of a second to anticipate or avoid a blow. But to move forward in pursuit, to cover ground against a fast boxer who knew how to run, he had to hop.

Parkhurst remembered from his reading of the habits of the Australian kangaroo that when grazing he went down on all fours on his short forepaws, his rear elevated by the long hindquarters and in this manner he could inch forward. But when alarmed and needing speed, he would stand erect and cover twenty feet at a hop. Since the ring was twenty-feet square, a leap could take Matilda from one side to the other. He was able to cut this down to half, but less was difficult for him; more, might cause him to overreach himself. Towards the end of the round Parkhurst was certain that Dockerty was timing Matilda's bounds and managing to be elsewhere. Just at the bell Matilda was on one side of the ring and Dockerty clean on the other, with the referee in the middle. A roar of laughter shook the arena and there was applause from the front sections.

The referee went over to Dockerty's corner. There was too much noise to hear what he said, but it was plain from his gestures that he was warning Dockerty that unless he did some fighting, he was liable to disqualification. This set off a jawing match between the referee and Pinky Schwab, with Pinky obviously telling the referee to go peddle his apples somewhere else.

Matilda was still refusing to sit and as Billy Baker kept bellowing into the trumpet of his ear, shook his head almost angrily as though to shut off the sounds that were preventing him from concentrating. He looked nervous and Murray O'Farrell dictated that he definitely seemed to have lost some of his cool. To Parkhurst the boxing writer yelled, 'But what's he going to do? What's he thinking?'

'He doesn't think,' Parkhurst replied. 'He isn't a person, he's a kangaroo. If Dockerty goes near him, he'll zap him. If he doesn't and this goes on for fifteen rounds, you and I will have to leave town.' He called up to Patrick Aloysius, 'Tell him to stand still. He's crazy to chase him. Tell him to stand still and if Dockerty doesn't come to him, the referee will make him.'

Ahearn was climbing down the steps as the klaxon warning seconds to leave the ring had sounded. He threw Parkhurst a look of utter disgust and said, 'You tell him. He no spik'a da English.'

At the bell a great roar arose from the crowd. It was the beginning of round three and up to then none of Matilda's opponents had lasted more than two rounds with him. Dockerty had already scored some kind of moral triumph. The crowd was with him. They wished him, if he could not win, at least to survive, to achieve even this small victory.

Ambivalence was tearing Parkhurst apart. He was seeing Dockerty in a different light. He was grateful to him for surviving; he wanted Dockerty to go on, to give the fight fans what they had paid for, drama, shock, the unforeseen, resulting from the head-on clash of two determined spiritual and physical forces. He also desperately wanted Matilda to win.

The fighting brain of Matilda had apparently suggested a

251

means of solving the impasse by a method that had worked once before. At the peal of the gong that initiated the third canto, he bounded diagonally across with one leap and was waiting when Dockerty got up off his stool and dropped him in his corner with a short right uppercut for a count of eight. Seventy-five out of more than a hundred correspondents used the word 'pandemonium' to describe what broke loose.

Pinky Schwab was up onto the top step screaming, 'Foul' claiming that Matilda had hit Dockerty before his man was on his feet, but the referee, after wiping the champion's gloves and looking into his eyes, ignored the protest and motioned him to go in and fight. Somebody pulled Schwab back down again.

Dockerty attempted to clinch but was held off with three straight lefts and a right to the ribs that made his legs sag.

The crowd was frantically begging Dockerty to stay away, but he was hurt and Matilda herded him into a corner where Lee covered up trying to hide behind his elbows, until Matilda brought his guard down with a left hook to the body and then chopped him to the canvas again with a short right.

Lee was able to regain his feet at the count of nine, glassy-eyed, but managed to grab his antagonist and hang on. The referee was slow in getting over and they clung there like a couple of lovers spooning in the park. Matilda always disdained in-fighting in a clinch, either as something inefficient, unpretty or both. His success was filling his heart once more with love for his opponent and he rewarded his efforts with the usual facial.

The referee pried Dockerty loose but he immediately fell into another clinch and when he tried to work on the body, Matilda tied him up and they wrestled all over the ring. The crowd knew that Dockerty was fighting for his life, was on its feet and standing on the benches.

On the break, Lee tried a sneak punch but Matilda slipped inside and zapped him with a lightning right and down went Dockerty again. This time it did not look as though he could possibly be getting up. But at the count of seven the bell rang. Had the round lasted three seconds more, Matilda would have been undisputed middleweight champion of the world.

Pinky and Dockerty's handlers swarmed into the ring,

dragged him up onto his stool and applied restoratives. It was obvious that for this fight Lee had worked himself into superb condition for he recovered with remarkable rapidity and pushed the bottle of smelling salts away when the referee and the club doctor came over to look at him to weigh his ability to continue. He was on his feet waving them away, calling them obscenities and shouting hysterically that he would kill the son-of-a-bitch in the next round.

The klaxon sounded for seconds to quit the ring. Parkhurst was digging his fingers into O'Farrell's shoulder, his voice breaking with excitement, 'It's all over! That's the punch he hit him with that first time I saw him in Camauga. Dockerty won't get up the next time!'

At the bell Dockerty was off his stool and managed to avoid Matilda's rush as his seconds screamed, 'Keep away! Keep away!' But instead, he went completely berserk and drawing strength from adrenalins of rage and humiliation, he began to pour blows in upon Matilda, left and right swings, jabs, uppercuts, looping rights, in a rally that brought forth such a Niagara of sound from the hysterical crowd that the earth shook beneath its feet.

For at the same time as Dockerty's rally was rousing them to the highest pitch of excitement, they were seeing the most extraordinary exhibition of defense as Matilda picked off punches, sometimes almost before they had started, blocked, slipped, pulled back, turned just enough to let a murderous hook go by, smothered the attack, occasionally pausing only to cuff Dockerty about the ears. There was an expression of great contentment on his face. For the first time he was enjoying himself, and when at last, exhausted, Dockerty was compelled to clinch, he gave him that last loving lick that brought a cry of triumph from The Bermondsey Kid at the foot of his corner, causing him to throw up his arms and shout, 'That's it! That's all! He's given 'im the kiss of death.'

But it was not yet. Fate, the dramatist, whose vagrant moving finger was inditing the script, had something else in mind. For as the referee slapped Dockerty's gloves down from around Matilda's shoulders, calling 'Break!' and the kangaroo dutifully stepped back, Lee had a moment of sanity and the pleading cries from his corner to, 'Keep away!

Keep away! Keep away! Run, Lee!' reached his ears. He once more proceeded to do so, though on shaky legs, but nevertheless obliging Matilda to take up the pursuit once more.

But this time more than sanity returned to the champion as he remembered his earlier analysis of Matilda's weakness in chase. He began to exploit this by sharply changing angles of movement as he back-pedalled, so that to catch up with him Matilda was in a sense compelled to all the moves of the various pieces on a chess board—pawn, queen, bishop, tower and most awkward of all, the knight's move, as Dockerty abruptly shifted direction. Two minutes of the round had passed when the pursuer was again forced into this manoeuvre and this time Dockerty was waiting for him. And, as Matilda arrived slightly off balance, his left reaching, his right cocked for the finisher, the tired man managed to swing two punches into Matilda's body. It was the first time that he had connected solidly and to his own astonishment, so much so that instead of following up he stepped back prepared to resume his retreat to survival.

It had all transpired so quickly that the crowd was not yet aware of what had happened. Only the experts appraised the extraordinary results and Parkhurst, staring at Matilda, was shaking O'Farrell by the shoulder as he cried, 'My God! Look at Matilda! What's gone wrong? What's happened to him?

For the kangaroo too, had backed away and now was standing over by the ropes with both his stubby arms wrapped around his middle and a look of injured surprise on his face. And, as they watched, they saw two large tears issue from the animal's eyes and run a dark furrow down the side of his cheeks. The expression on his face, thought Parkhurst, was that of a child who for the first time has been struck by an adult, and simultaneously suffered both pain and disillusionment.

Parkhurst saw more than that. At the bottom of the steps of Matilda's corner, The Bermondsey Kid had seized a towel saying, 'Oh, my Gawd! That's done it. The poor bloody barstid! He's finished!' and prepared to throw it into the ring. But both Bimmie and Ahearn grabbed his arm and prevented him.

'Not yet, you idiot!' Ahearn cried. 'Give him a chance. He might still have a zap left in him.'

The referee motioned for Dockerty and Matilda to resume fighting again. But Matilda stayed where he was, his back to the ropes, the tears now streaming down his face and for a moment the crowd, uncertain what had happened or what was about to happen, fell into a hush. Parkhurst heard Billy Baker's cry of anguish, 'Don't let him hit him again! He's been me pal!'

For an instant the scene was as static as though a film had been stopped in its race through the projector to flash only one still frame upon the screen. Some instinct told Dockerty that mysteriously the tide had turned and with a return of his battle rage, too compulsive to consider caution or that Matilda might be trying to lure him in, he sidled close.

Matilda's arms were still wrapped ruefully about his middle and he made no attempt to defend himself.

Dockerty, summoning his last strength, whaled him again in both flanks with a left and right hook.

Matilda dropped down onto his forelegs, hunkered in front of his attacker and stayed there.

A roar, a shout, a burst of sound to dwarf all previous cheers broke from the mob. Parkhurst shouted, 'My God, he's grovelling!'

O'Farrell yelled, 'Come on, you yellow bastard! You ain't been hurt!'

But Frank Petrie, who was sitting next to him, laughed and said, 'There goes Matilda. Who would have guessed it? He can't take it down there.'

And Matilda continued to cringe at Lee Dockerty's feet, as meek as a kitten, while the champion stood over him screaming obscenities and begging him to get up so that he could have another shot at him.

For a split second it was one of those dangerous moments when everything could go out of control and the crowd turn into a mob and break loose. Nobody knew exactly what had happened; how much steam Dockerty's punches had carried; how badly hurt Matilda was; whether he had been knocked down, fallen down, or was quitting cold.

To add to the critical instant the tornado storm of sound contained a massive jangling of instruments to a chorus of boos, as the hippies once more disillusioned, abandoned their hero.

Even the veteran referee was confused for a second as to what could be considered 'down' with a boxing kangaroo. But his gloves were touching the ring floor, even though he looked neither hurt nor physically distressed, but only grieved beyond description. The referee quickly regained his head, turned Lee Dockerty around, spinning him by his shoulder and pushing him towards a neutral corner. The discipline of a hundred fights took over and the champion, now himself weeping hysterically, stumbled off, holding himself up by the ropes. The veterinary doctor of the S.P.C.A. was running around the ring waving and shouting something, but no one paid any attention to him.

The referee about-faced to see whether Matilda had used the time that Dockerty had given him to get up. The knockdown timekeeper had kept his head and held his hammer until Dockerty obeyed instructions. Matilda was still there on all fours. The gray-clad arm of the referee thereupon moved rhythmically up and down ten times, followed by the palm spread downwards signal of ten-and-out.

Torrents, seas, oceans of cheering broke over the ringside. Dockerty's handlers were in the enclosure hoisting him high in the air in a maniacal victory dance. The bell clanged and clanged unheard. Billy Baker crawled into the ring, went over to Matilda, kneeled down beside him cradling the melancholy head in his arms, fondling, stroking and murmuring to him. And thereafter, still on all fours, Matilda suffered himself to be led to his corner, his gloves removed and his chain and collar replaced.

Police congregated around the ringside to keep delirious fans at bay. The bell finally overcame the cheering and the announcer was able to give the statistics: 'Time, one minute and twenty-nine seconds of the fourth round. Winner by a knock-out and still champion—Lee Dockerty!'

Microphones were thrust in front of Dockerty, cameramen crowded. A police captain with gold braid on his cap signalled to Matilda's corner and said, 'Get him out of here. We don't want any trouble.'

But the S.P.C.A. man who had managed to climb through the ropes was making a cursory examination of the animal. Only Parkhurst actually knew who he was and shouted up, 'Is he all right?'

The veterinary, whose expert hands had been sounding Matilda's body without drawing any reaction of pain from the kangaroo, nodded and replied, 'He's okay. He doesn't seem to be hurt.'

The police captain again signalled for them to leave the ring and as they did so, nobody even noticed them in the excitement of Dockerty's triumph.

Matilda inched through the bottom rope and down the steps on all fours with Baker leading him. The little Cockney ex-fighter was crying as they took off along the aisle towards the dressing-room. Parkhurst looked up as Ahearn and Bimmie followed. To his utter astonishment the two managers were grinning as they hustled themselves out.

Parkhurst experienced a desperately sinking feeling at the pit of his stomach, as he watched them go. Somewhere, somehow things were not right. A feeling at the nape of his neck and goose flesh on his arms warned him that something about this dramatic and yet in a sense ridiculous ending was fishy. He could only think: *Those two bastards, once again, know something I don't,* before he turned to his telegrapher to dictate his story of the débâcle.

# XXIV

THE PARKHURSTS SAT at their breakfast table the following morning, Duke mumbling and grumbling over his copy of the late edition of the *Mercury* with its horrifying, front-page smash picture of Matilda down on all fours at the feet of Lee Dockerty; Birdie upset for him because he seemed to be upset, though deep down inside of her she was deliciously and deliriously happy.

For he had informed her several nights before that by official fiat of the Federal Bureau of Investigation she was henceforth to be known as Mrs Duke Parkhurst, with a belated wedding celebration to be held—of course—at Luchows. She did not understand what the F.B.I. had had to do with it but what was clear was that their secret marriage was to be made public to everybody, oh everybody in New York, and hence all the world, would know that Birdie McBride was Mrs Duke Parkhurst. And, therefore, having prepared him a good breakfast—like all troupers she had learned to cook the hard way—she sat opposite and worshipped him.

Parkhurst read not only his own account of the fight but those in the other morning papers and here at least he had reason for some satisfaction, for the majority of them concentrated upon Lee Dockerty's great comeback from the brink of defeat, the drama of the battle, the phenomenon of one of the greatest crowds to witness a boxing match and praising the champion's courage and stamina to the skies. Two of them in their various papers started a serial life story of Dockerty. Only that old fox of *The Times*, Cassius Jones, had a sniff at the astonishingly sudden collapse of Matilda but did not dewell upon it too greatly, taking the view that a critic would have to know something about the internal anatomy of a kangaroo before attempting to render judgement.

A three-million-dollar gate, of which close on two million would go to feed undernourished children and give them a chance in life, was a triumph indeed and yet Parkhurst was unable to savor it. He felt sad, mystified, somehow humiliated, suspicious, anxious, sick at heart and very angry with himself. His story, in the same vein as those of the other boxing writers, was adequate, well written, except that since it was a *Mercury* promotion his prose was slightly more lush. It gave the tale of the struggle in detail with all the sidelights, the crowd reaction, the by-play in the corners, the curve of his narrative rising to the stunning dénouement; thrill following upon thrill with, like the others, tribute to the game boy who had got up off the floor to win and thereby had proved himself.

And Parkhurst knew as he scanned his own lines, his mouth beneath his moustache curling with disgust, that what he had written was a lie and in particular his account of the ending. The trouble was he did not know where the lie lay. He only knew that he had not told the real story because he did not know what the real story was and it burned him. The Matilda saga had been his from the very beginning; his discovery, his inception, his development, his bringing it to its incredible fruition, and then to find himself left with lingering doubts and dissatisfaction! There must be an answer. He would have those three in his office, lock the doors and if necessary beat it out of them.

Birdie asked, 'Why do you love me, Duke?'

It brought him to with a jolt and he dropped the copy of the *Mercury* and found himself looking into her adoring eyes. He said, 'I don't know. I just do, I suppose.'

She said, 'You oughtn't to, because I know all about me. But you're like a being from another world. I love you so it makes my tummy hurt. I worship you and look up to you and want to hug and hold you tight every minute. You know everything.'

259

For the moment she had banished all the megrims that had been haunting him and he said fondly, 'Love, I'm afraid, is one of the things I know least about. But I do know that every time I look at you, I'm overwhelmed by tenderness. Somewhere you've touched my heart, Birdie, and it will last. It will never change. That's the unanswerable mystery about it.'

But the word 'mystery' put him back upon the paths of darkness again and simultaneously Birdie said, 'You're worried, aren't you?'

He finished his coffee, wiped his lips and arose. 'Yes, I am,' he acknowledged.

'Why? Wasn't it a wonderful success?'

'Financially, yes. Morally . . . ? And just to prove to you that I don't know everything, I have no idea why Matilda sat down on his fat bum, or rather forepaws, after being hit a couple of punches in the flank by a fighter so badly hurt himself he could hardly stand on his feet and barely lift his arms. Birdie, do you suppose that somehow the great Parkhurst has made the biggest boob of the century out of his clever self?'

'Oh, Duke!' Birdie was horrified. 'You couldn't! And anyway, whatever happened to Matilda it wasn't your fault.'

Parkhurst sighed and said, 'You wouldn't understand, Birdie,' and as always had that tiny vestige of satisfaction that this was so.

Birdie said, 'Perhaps it was the first time anyone ever really hit Matilda.'

Parkhurst paused at her chair and dropped one of his big, ham hands upon her delicate shoulder. Smiling down upon her he said, 'After all those years of exhibitions and taking on all-comers and all his fights? And you want to know why I love you, Birdie?' He bent down and kissed her with that ever recurring knowledge that he would cherish her forever and went off to the *Mercury* building. By the time he got there the deep sense of foreboding had returned.

Duke Parkhurst sat in his office awaiting the arrival of Bimmie, Paddy Aloysius Ahearn and The Bermondsey Kid. On his desk before him was a telegram from the owner of the paper, 'CONGRATULATIONS DUKE AM AWARDING YOU BONUS

Parkhurst pushed it aside. It was ashes. Another time it would have filled him with pleasure and perhaps the thought of a new car. Now he did not want it. He would bank it for Birdie, since Uncle Nono no longer would be underwriting her.

Uncle Nono! Had the tables been turned? Was the mysterious avuncular shadow now laughing at him? Had he managed somehow to slip something into Matilda's soup? No! The kangaroo had been in perfect condition when he had entered the ring. Then if it had been Uncle Nono or one of his men, how had he scored?

He remembered that he had been tempted to seat Birdie in the Donor Section on the chance that she might see or identify the man she knew only as Joe. And then he had decided that this would be both cruel and unfair to Birdie. For her, Uncle Nono, if indeed it had been he, was a closed chapter. So he had located her some thirty rows back in the arena and because he was that kind of honest, had made out his check for a hundred dollars to the *Mercury* Free Food Fund for Hungry Children.

He also realized that the identification or exposure of the Cosa Nostra's bosses in New York was not his can of peas to open. Let Clay on the city side, the cops and Feds worry about that one. But it was his responsibility to ascertain why the first time Matilda had been struck a solid punch, he had quit cold, and his greatest twinge of conscience came from the fact that he had not said so in his account of the battle. It was only Cassius Jones who had suggested he thought that Matilda might have got up but that on the other hand, those first two blows to the body could have taken all the fight out of him which, in the final analysis, was a legitimate purpose of the exercise in a boxing match. To force your opponent to quit was just as effective as rendering him unconscious or scoring more points over fifteen rounds.

Reading over this sentence again, Parkhurst murmured, 'He's right. He quit, he quit, he quit! But why? One more zap and Dockerty wouldn't have got up again.'

The sports department copy boy stuck his head inside the

door and said, 'There are some guys outside to see you.'

Parkhurst said, 'Send them in.' He swept the newspapers together, with their damning photographs of Matilda hunkered down on the canvas on all fours, like a Mandarin kow-towing to the Emperor and when the three men entered he was leaning back in his swivel chair, playing with a paper-knife. He said, 'Come in, fellas. Shut the door behind you and sit down.'

There were three chairs ranged in a semi-circle about his desk and the visitors settled themselves gingerly upon the edge of each.

Nobody said anything for the space of almost a minute, while Parkhurst regarded them heavily and morosely. He finally said, 'Okay, so what happened?'

The three men looked at one another uneasily to see who would be the one to speak and answer the question. None of them wanted to do so. Eventually Bimmie, who was in the middle under the anxious glances of his two partners said, 'Well see, Mr Pockhurst, it was like you wrote in the paper. Dockerty kept coming up off the floor and Matilda got careless for a moment on account of all that noise and yelling, see? And he forgot to get his hands up, and Lee got in two good ones which would have knock out anybody. The kanagaroo couldn't get his wind back. The punches was low. Lee fouled Matilda. The ref should of disqualified Lee for fouling. He would of got up in another minute, except the referee was giving him a fast count . . .'

Duke Parkhurst contemplated the thin blade of the paper knife and said, 'Horse feathers!'

Bimmie, whose layers of blond hair had risen to half-mast, looked over for help to Patrick Aloysius Ahearn who had been masticating his tobacco, his face as usual sardonic. He glanced at either side of Parkhurst's desk for a spittoon to shoot at.

Parkhurst said, 'There isn't any. Swallow it. What happened?'

Ahearn's eyes bulged for a moment. The adam's apple in the wattles of his throat registered outrage as the tobacco juice went by. Then, shifting his cud into his left cheek, he said, 'Matilda punched himself out. He didn't know how to

262

pace himself. You've seen that happen. He'd never had to go more than two rounds. He gave everything he had in that third round and had Dockerty out cold, if he hadn't been saved by the bell. Three seconds more and we win. He didn't have nothing left in the fourth. He'd given everything he had. What more did you want out of him? After Dockerty hit him that low punch, he couldn't raise his arms any more. The referee should have...'

'Hogwash!' Parkhurst interjected and beat a little tattoo on his desk with the knife. 'He was as fresh as a daisy. And the punch wasn't low either. Birdie said she thought maybe it was the first time he'd ever been hit, but that's crazy. I don't know anything about kangaroos, but it looked to me like he quit cold.' He turned and pointed the blade at The Bermondsey Kid and said, 'Now you tell it like it was, Billy.'

The three men exchanged panic looks. Bimmie and Patrick Aloysius both had their lips formed to let words issue. Parkhurst said, 'Shut up, both of you. I'm asking Baker. All right, Billy, come clean. What happened?'

The Bermondsey Kid blinked, avoided the pleading gazes of his two companions and replied, 'You've about got it straight, sir—Miss McBride, I mean. That's what 'appened.'

'What? Straight of what?'

'Like you've just said it, sir. It was the first time ever that Matilda's been hit a hard punch in anger.'

Parkhurst's paper knife fell on the desk with a metallic clatter and he whipped himself erect on his swivel chair crying, 'What? Are you kidding? In eight years, or however long you've had him, the first time ever? But that's impossible, I don't get it. Are you trying to tell me...?'

Baker's wrinkled, bashed-in face was a picture of misery but as nothing compared to the horror settling upon the countenances of Ahearn and Bimmie as the trainer turned upon both of them and said, 'Ow, what's the use of tryin' to lie? The gentleman ain't no fool.' He turned back to Parkhurst and said, 'You see, sir, you can have an act with a kangaroo like I had in the circus, but you can't hit him. Slug a kangaroo and he's finished, he'll never box again. I've never said nothing about this, but I had one I trained three years before I found Matilda. We was a-shiftin' of him one day, loadin' up to travel, and I says to the roustabout, "Here, hold his chain a minute, while I look to his cage." The animal, not knowing him, got nervous or frightened or something, and hit the bloke a clout in the ear and the bloke hauled off and hit him back a blowser. That done it. The beast was finished; he'd never box again. Three years gone to waste. It wasn't the bloke's fault, he just hit him back instinctive like. But you can't do that with a kangaroo. Once he's hurt...'

Alarm bells were beginning to go off within Parkhurst's head and he hoped that none of the three would see the pinpoints of moisture he felt were forming at the base of his moustache and in the bushes of his eyebrows. He said, 'But I thought you said they boxed in their wild state—with each other? That you were merely using something they did naturally, to make a circus act?'

'That's right, sir, they do,' Baker agreed. 'But it ain't boxing like we know it. It's either over a lady friend to see who gets her, or to find out who's boss of the 'erd. The one that gets in the first good wallop wins. The other one will quit and what's more, he'll never fight again. That's your 'roo for yer. He's got to think he's top dog, see? He's a lot like some humans in that. As long as he thinks he's the greatest, he'll knock your brynes out. But once he finds out there's more to the game than just dishing it out, he folds up like an umberella.'

Parkhurst wondered whether they could feel the fall in temperature caused by the cold that had gathered at the pit of his stomach. He queried, 'Any kangaroo? All kangaroos? But I thought Matilda was special, and could box like a streak? And, by God, he could! I saw it with my own eyes and so did everybody else.'

'That's right, sir,' Baker said. 'He was, and he could. He was special, like I told Bimmie 'ere from the very beginning. He *was* different, sir, the way he took to the game like no other kangaroo I ever saw. But that don't change 'is nature, do it, sir? I never hit him in all the time I was teachin' or we was sparring together, or in the act. And it was only when he got to be so expert I could hardly handle him meself that I could tyke a chance on lettin' him meet some of those

264

jackaroos at the fairs. They hadn't an earthly with him, no more than a two-year-old kiddie would. You could see their swings comin' a mile off. They'd hardly get set when he'd feint 'em, let go a one-two and Bob's your uncle. And, I'll tell you another thing, sir,' Baker added.

'Yes?' said Parkhurst.

'If I'd know'd who Lee Dockerty was when he come in under his real name, down there in Camauga, I'd never have let him get in the ring with Matilda in a million years. I'd have give him the five hundred dollars. See, you know how it is, with a pro. Unless you stop him with the first punch, he's bound to hit you a couple of times. Even a genie like Matilda. I used to keep me eye open for any pros that might try to sneak in. You know, a cauliflower ear, a bashed-in nose or maybe an eyebrow cut up, and I'd make some excuse and bring on the stooges. Dockerty didn't have a mark on him, so he got by and we was lucky. He'd had a couple a drinks under his belt and Matilda nailed him early.

'I can tell you, sir, I nearly had a fit when I read in your article that Matilda had knocked out the middleweight champion of the world. Matilda was me bread and butter. He was me retirement and old age insurance. He was the pub I was going to buy back in Bermondsey. If Dockerty had clipped him, that would have been the end.'

Added to the cold sweat now flowing from beneath his armpits, Parkhurst felt the room turning about him as well. He clutched the side of his desk, fixing his eyes upon the famous photograph of the Dempsey-Firpo fight on the opposite wall, to make the gyration stop. He said, 'I see. So you never let him into the ring with a pro? What about all those build-up fights?'

Baker did not reply. The silence of the three men gazing down at their feet and twiddling their hats, no longer even daring to look at one another, began in Parkhurst's reeling head to take on the noise of a thunderclap. He already knew, but he had to hear it. He said, 'Okay, you three crooked sons-of-bitches, what did you do?'

Bimmie appeared to recover from his terror first as an idea struck him and said, 'Nothing, Mr Pockhurst, nothing that wasn't on the up-and-up. See, it was just like making up

265

an ack for a show. Billy and his boxing kanagaroo was a ack I booked into Matson's Monster Carnival. See, Matilda was boxing Mr Baker here and it looked like they was trying to knock each other for a loop every night but it was an ack. So when we got those other fellas to go along with Matilda, it was just a part of the ack, wasn't it? We never said it was anything but an ack, did we? Because who would get a crazy idea that a kanagaroo could be a middleweight boxing champion of the world?' He choked as he realized what he had said referring to his hero and benefactor and recovering apologized.

'Begging your pardon, Mr Pockhurst, and not meaning any offense but you started it with your article, when you wrote that for you and the *Daily* and *Sunday Mercury*, Matilda was the middleweight champion of the world. I guess I would have not have known my job very well, booking attractions, if we hadn't made it a part of the ack then, after all that publicity you given and said like he was the greatest boxer in the world, and if he could knock out Lee Dockerty, the middleweight champion, then he was the champion and not Dockerty. And then when you told me, like, how now if Matilda got a good record and could fight his way up to a return match, we could make maybe a million dollars . . .'

Parkhurst felt the net closing in about him. He regarded the three bitterly and said, 'So you mean to tell me that every goddam fight Matilda had, from that day in Camauga until last night, was rigged? Why you dirty, lousy, thieving bunch of . . .'

Patrick Aloysius took another swallow of his tobacco juice, coughed and let words as harsh and bitter as Parkhurst's dribble out of the side of his mouth, 'Come on, Duke, don't give us that! What did you think Dockerty's build-up to get the title fight with Cyclone Roberts was? Out of fourteen, five were on the level and three of those were with cripples, one of whom died two weeks after Dockerty knocked him out. Between Pinky and Uncle Nono's man Marcanti, they rigged nine elegant splasheroos. Everybody knew it and so did you. But you, my friend, weren't too proud to use Lee Dockerty for your big charity show, were you? If you were so goddam holy, you'd have refused to use a boxer whose record stank like Dockerty's. You thought you were pretty smart, taking a fall out of Pinky, Lee and Uncle Nono in a way they couldn't hit back, but yours wasn't the first gag like that to backfire, my noble friend.'

The Bermondsey Kid spoke up. He said, 'You oughtn't to be angry with Mr Bimstein and Patrick Aloysius, Mr Parkhurst. It was all my fault. See, they didn't know about Matilda.'

Parkhurst was almost glad to be able to drag his eyes away from the cynical gaze of Ahearn. He said, 'Know what about Matilda?'

'About what I've told you, that once a kangaroo was hit and hurt, he wouldn't box no more. After your article appeared they came down south and caught up with us at somewhere in Oklahoma, and said you was backing Matilda for champion of the world, and we should take on all-comers to build up for a million-dollar return match with Dockerty. "And get me kangaroo ruined?" I says. "Not ruddy likely!" I explained matters, how somewhere along the line somebody was a sure thing to hit me pal a chop on the whiskers, or bloody his nose, and that would be the bleedin' end of Matilda, the world's champion boxing kangaroo.'

The mists surrounding the affair began to lift, though what they showed Parkhurst made him no happier. He said, 'So you formed a partnership?'

The Bermondsey Kid nodded. He said, 'Yes, sir. After they swore they wouldn't let Matilda be hurt, I made 'em swear it on the bloomin' Bible.'

'And naturally as a good Catholic . . .' Ahearn said.

Parkhurst was beginning to marvel now at the colossal effrontery of the unholy three. He said, 'And Patrick Aloysius, of course, made all the arrangements. But how the hell did you get them all to agree not to hit Matilda? And how did you know you wouldn't be double crossed?'

A gleam of pride came into Ahearn's eyes. 'Human greed,' he said. 'You've got to be big. We gave them what they asked for. What the hell? We weren't interested in peanuts, we were shooting for the jackpot. A couple of times when we ran into trouble with some of nature's noblemen who placed honor above self, I used two of my own stable, out of town, under assumed names.' For a moment his face split in a sideways grin showing tobacco-stained teeth and he continued, 'You remember we had a little trouble with some of Uncle Nono's boys along the way. But that Joe Marcanti and his goons never caught onto the fact that we were buying up the opposition.'

Parkhurst said, 'You mean they were that dumb that they never caught on, having done it themselves?'

Billy Baker said, 'After the way Matilda zapped Lee Dockerty and how he was going, how would they twig unless they knowed the secret of 'roos? Only a circus man would know you can't hit a 'roo.'

'They could have outbid us if they had,' Ahearn continued, 'but instead they went through the usual routine of intimidation and stuff like Evil-Eye Finkel. I suppose that's why maybe they had to lug Joe Marcanti's head back across three State lines to find out if it would fit a body. I see by this morning's papers that it did.'

'So you see,' Bimmie put in, 'we was just looking after Matilda by building up the ack: Bimmie, Patrick Aloysius, The Bermondsey Kid and their World Championship Boxing Kanagaroo. I wouldn't do anything dishonest, Mr Pockhurst. But see, it's like when you go the the theater, and say the show is about how one of the actors picks up a gun and shoots another actor. You wouldn't expect that gun to be loaded, would you, Mr Pockhurst? So every time he shoots, you got to get a new actor? So instead he puts a blank in and the actor falls down and pretends he's dead. Everybody in the audience knows he ain't dead but they feel like he is. Is that dishonest, Mr Pockhurst, I'm asking you?'

There was something so fiendishly illogically logical in what the little manager with the wavy hair was saying that for a moment Parkhurst felt his own ethics somehow pushed off center. He had thought originally of asking Clay to be present at this interview. Now he was glad that he had not. Fascinated, his hands folded, he leaned forward on his desk and said, 'So you come up to the championship fight at Jericho Stadium with a record of fourteen phoney knock-outs.'

Billy Baker protested in a slightly injured tone, 'Oh no, Mr Park'urst, there weren't nothing phoney about them knock-outs. Matilda 'e bloody well zapped 'em. The arrangement was just that they wasn't at any time to 'it '*im*. If they could stand him off for eight or ten rounds that was all right with us.'

'So you see,' put in Bimmie triumphantly, 'there wasn't nothing dishonest, Mr Pockhurst. The knock-outs were on the level.'

Parkhurst said, 'Including the one in Chicago on CTS television, seen by ten million people?'

Bimmie expostulated, 'Mr Pockhurst, that was the most honest of all! We had a script for that one, you know, like I said in the theater— a regular play. And this fellow Cyclone Roberts was hired to play the part...'

'Of he who gets zapped,' Patrick Aloysius concluded, 'It looked good, didn't it? The sponsor, Mr Van Houven sold two million more pairs of jock straps on the strength of it. We ought to be getting a medal for philanthropy and you, my friend,' and here he pointed a grimy fingernail at Parkhurst, 'ought to pay for having it struck.'

In spite of himself, Parkhurst had to smile. He said, 'Okay, I've got the picture. And now we come to what in the old days was called the sixty-four-dollar question: Lee Dockerty was not going to be drunk this time. He was fighting for his title; he was going to be in condition and what's more, out for revenge. Sooner or later, unless he got clipped early, he was going to hit Matilda and hit him hard. What happened when you propositioned him?'

Once more a silence fell upon the three and the quick, fearful, guilty exchange of looks. With a pang that brought back all his doubts and discomforts, Parkhurst felt that the tale was not yet done.

Finally Bimmie, with the layers of his hair now three-quarters upstanding, looked down at his feet and replied, 'He said, "No".'

Patrick Aloysius threw back his silver head and laughed so that his jowls quivered. 'He said more than that. He said we could all take a flying you-know-what for ourselves. He said he was going to belt that effing kangaroo on his ugly mug, and knock him all the way back to Australia.'

Parkhurst asked out of curiosity more than anything else, 'What did you offer him?'

'Everything we had,' Ahearn replied. 'This was the big one we were all in for. Once we had the title for keeps, we could have got it all back.'

'I even threw in the money I'd saved for me pub in Bermondsey,' Billy Baker added.

And Paddy Aloysius mocked, 'My box at Ascot, and my villa in Monte Carlo.'

'My marriage, like, with Hannah,' Bimmie said. 'See... Well, she kind of pinned me down last year and then gave me the air. I thought maybe if I could show Mr Lebensraum three or four hundred grand in the bank so he should see his daughter wasn't marrying a bum, maybe she would take me back. We put it all in the kitty and offered it to Dockerty.'

'Who to your great surprise suddenly turned square,' Parkhurst noted.

Bimmie nodded and added, 'Like Patrick Aloysius here said, he told us what we could do.'

'And what's more,' put in Billy Baker, ''e told us what 'e was going to do.'

Parkhurst nodded and said, 'So you figured that with Lee Dockerty in that kind of a mood and profiting from his first experience with Matilda, he was a sure thing to get in at least a couple of licks and that would be the finish of it. The 'roo would curl up like he did and quit like a dog.'

'Not like a dog,' The Bermondsey Kid protested, 'like a kangaroo. See, it's the nyture of the h'animal.'

'That's right,' Ahearn nodded. 'Lee couldn't miss, could he? Unless he'd have been nailed early in the fight, like he was. Only he was in shape and got up. Even a Boxing wonder like Matilda can't block 'em all. And when i found out he had a china chin and a yellow streak, we had to protect ourselves, didn't we?'

The Bermondsey Kid protested again, 'He wasn't yellow! Like I said, it was just 'is nature.'

Parkhurst said testily, 'Yes, yes, yes! I know all that now.' He realized that they were stalling. 'What I want to know is

what you did when Lee told you to go jump into the river.'
He turned to Ahearn and said, 'And what the hell were you
and Bimmie grinning about when you climbed down out of
the ring, after Matilda had succumbed to a tap on the ribs
that wouldn't have upset a three-year-old?'

Silence again and the panic looks. Parkhurst's temper
suddenly got the better of him and he banged his ham-sized
fist onto the desk top so that the paper-cutter, ruler, paste
pot, inkwell and calendar jumped. 'Goddamit!' he shouted, 'I
want the truth! The whole truth...'

'...And nothing but the truth,' Patrick Aloysius
concluded for him, his face once more a mask of utter
unconcerned cynicism. 'Since you ask us so prettily, we
unloaded him on Uncle Nono.'

# XXV

DUKE PARKHURST'S ANGUISHED cry of 'What?' penetrated the walls and glass door of his office, causing the sportswriters to look up from their work and wonder what the hell was going on inside there with the Old Man.

Within, Parkhurst, his voice fallen to a whisper of sheer horror, repeated, 'What? You un...' And then, 'Before or after the fight?'

Patrick Aloysius said, 'What do you think? After the fight Matilda wasn't going to be worth a 1970 dollar bill, except to a zoo. See, with those big shots with an ego, if you will forgive the expression, like Uncle Nono's, they can't bear to be on the losing side. Off Matilda's record and sending scouts to watch him work out, he figures Matilda couldn't lose. So he had to have him.'

'He come to us with the offer,' Bimmie said virtuously. 'It wasn't us that went to 'im. That wouldn't have been honest.'

'I didn't want to sell,' The Bermondsey Kid put in. 'Matilda was me pal. He loved me like a brother. I was outvoted.'

Parkhurst asked, 'What was the price?'

Patrick Aloysius Ahearn eyed the sportswriter mockingly out of his drawn-down, Bassett-hound lids, shifted his quid and said, 'A cool million, which we split three ways.'

'With Lee Dockerty thrown in,' Bimmie added. 'Pinky's out. I'm his manager now.'

Ahearn, out of the side of his mouth, corrected him, '*We're* his manager, remember?'

Even Parkhurst, who thought that in his years as a boxing

commentator, columnist and sports editor he had seen everything, was momentarily stunned by this revelation. In the fight racket, as it had been conducted for years, it was nothing for the manager of a champion to make a private deal with the business agents of the challenger that if the latter won, the dethroned title-holder was to share in a percentage of the new champion's next several fights, or until a return match could be arranged. A world championship was an asset as valuable, tangible and negotiable as a wheat or cotton future and not to be dealt with in an unbusinesslike manner.

Parkhurst asked, 'Who did you negotiate with—Uncle Nono direct? Who sent for you?'

Ahearn replied, 'Some guy by the name of Renato. He's got some kind of a brokerage office downtown but he was talking for old Uncle Nono all right. He had the class and he had the dough.'

Parkhurst marvelled, 'And you mean to say that Uncle Nono handed over a million bucks for Matilda in cash, with Lee Dockerty and good will thrown in? And now you've got the middleweight champion of the world and all Uncle Nono's got is a kangaroo with a memory of a bellyache and good for nothing but to rent out to model Ajax Webbing Reinforced jock straps?'

Patrick Aloysius said, 'That's what a college education does for you. They say on the Street that Uncle Nono was supposed to have gone to Harvard. His old man never should have let him near there. Now you take my old school, Hard-Knocks University...'

Parkhurst had not even heard the final sally. Falling back in his swivel chair and throwing up his arms, he was emitting shout after shout of laughter. Of all of the crimes in the book, this one perpetrated by Honest Bimmie and Company was the most divinely crooked of all.

While the three sat gloomily watching him, not knowing what to make of this slightly terrible mirth, the big man, his head thrown back, laughed and laughed and laughed as though he would never stop, with the tears streaming down from his eyes until wise old Ahearn, who in his day had seen men out of control, detected a note of hysteria in the laughter and said, 'I'll bet, if Uncle Nono's laughing, it's on the other side of his face.'

The sight of Parkhurst's tears suddenly brought some of his own to Billy Baker's eyes and he said, 'I've sold me kangaroo down the river to a murderer. He was like a brother to me. What will this Nono bloke do when he finds out Matilda won't ever lift a paw to fight again . . . shoot him? He was me pal.'

Bimmie said, 'Look, Billy, we've been all over this. Your share of the split, what with the movies and TV, is half a million dollars. You can buy up all the kanagaroos in Austerlia.'

Patrick Aloysius added, 'The deal's made, Duke, the money's in the bank. What are you going to do about it?'

The sheer insolence brought Parkhurst round. There were still tears in his eyes, but they were now from anger and not from laughter.

'Expose you bunch of bandits! Write the story! Tell the truth! I'm going to crucify the three of you. This whole business has been crooked from the beginning to the end. There hasn't been a single honest . . .'

It started Bimmie off on one of his favorite schmooses. 'Mr Pockhurst, what's honest today? Everybody got his ideas and rules, which is to try to make out so he shouldn't hurt the other fellow too much, especially when the other fellow's a dope and asking for it. It all depends on how you feel about it, ain't it? And whether or not anybody bothers to call the cops.'

Patrick Aloysius Ahearn suddenly interrupted, 'Shut up, Bimmie! I've got something to say.' He turned to Parkhurst and on his hang-dog face was an expression of insolent derision and his voice was mocking. He said, 'Okay, sweetheart, you want to talk about honesty? Now take that big amateur boxing tournament you run, which five thousand kids enter to get their brains knocked out for a medal with a chip diamond in it worth seventy-five bucks wholesale sanctioned by the Amateur Athletic Union, on account they get their cut in donations and their officials can walk around the ringside in hard hats and monkey suits, wearing badges. You take in a hundred grand at the gate and a couple of million dollars in publicity and extra advertising by reason of the pulling power of your paper. All based on conning dim-wits to fight for free. A lot you or your paper care what happens to them afterwards, when they turn pro and get their brains scrambled. You've got yours. A couple of months ago I had a kid come to me to get him a fight. He'd been in one of your Diamond Gloves six or seven years ago. I wouldn't send him into a ring. Punchy and half-blind at twenty-five. I gave him a stake and told him to go and get a job. But who wants a guy with a busted body and an addled brain? Like Bimmie said, what's honest today? You tell me.' And now he simply disregarded the fact that there was no gaboon and contemptuously fired a steam of tobacco juice at the foot of Parkhurst's desk.

Sick to his stomach as he was, Parkhurst found himself regarding Patrick Aloysius with dawning respect. He was a veteran scoundrel, educated in every phase of prize-ring rascality; he himself lived on renting out the skill and bodies of young men to be exhibited in the ring, but he had his own code of ethics and honesty and a certain likeable side. There were some things at which he would stick. He never overmatched his boys; he never let them stay in there for a beating when they were outclassed. He saw that they were properly trained and fed and when their usefulness as fighters was at an end, he found them jobs. None of Patrick Aloysius's boys were ever seen lurching about with the shuffling gait of the punch drunk, mumbling for hand-outs.

And he had struck home with his attack. The *Mercury's* gigantic annual amateur boxing tournament was one of the sports events of the year but deep down Parkhurst had never been happy about it. Patrick Aloysius had not used the actual words but everything he had said had caused it to loom in the back of Parkhurst's mind—exploitation of the simple-minded. Feed up hungry kids with the big charity and then destroy them later. Honest? That uncouth, ungracious, unappetizing ghetto boy used the word like a sword.

But he could not afford to let himself and his paper down before these three miscreants. He reiterated but with not quite so much assurance, 'I said I'll crucify you and I will!'

Patrick Aloysius merely leaned back in his chair, his lips curling, eyeing him derisively.

It was Bimmie who took the floor again, 'Look, Mr

Pockhurst, where is it against any rules if this here now Mr Nono wants to buy himself a boxing kanagaroo ack that's been famous all over the United States and Austeralia, and we got it to sell? Especially, like, now after we give our whole end of the gate to your charity. So he throws in Lee Dockerty in the bargain who is glad to get rid of a schmo like Pinky and go with some good managers. Okay, Mr Pockhurst, so you can do it. When you write your column, sure you could crucify us and make us out like a bunch of bums. But why crucify yourself along with it, all for a lousy kanagaroo?'

'What do you mean, crucify myself?' Parkhurst growled, 'I said . . .'

'Look, Mr Pockhurst,' Bimmie expostulated, 'you're a big shot in this town. Everybody got a lot of respeck for you and your writing about everybody and everything should be on the level, and nobody shouldn't do nothing crooked. And you done a great job for them kiddies. Off a million and a half every hungry kid in town can eat himself sick. So what are you going to do? Give all that money back to everybody that bought a ticket?'

Duke Parkhurst stared gloomily at Bimmie and said nothing, for the thought had already struck him in the midst of his indignation.

'See,' Bimmie continued, 'if you knowed that Matilda couldn't take a slap on the wrist, then you took in three million dollars under false pretenses, which like you would say, ain't honest. And if you didn't know it, you got to write yourself off in your story as a schmock, because a big promoter is supposed to know the kind of show he's putting on and you still got to give the money back for cheating the public. Not to mention, like Paddy Aloysius said, about Lee Dockerty's record.'

Duke Parkhurst felt a compelling sadness welling up within him. All he had wanted originally was to have a little fun with Pinky and Lee Dockerty and when it looked as though he had latched onto something, put up a show that would feed a hundred thousand children. The jaws of a trap had snapped shut on him.

Patrick Aloysius was aware of Parkhurst's dilemma. It had been like watching the champion who had been winning all the way, suddenly drilled by a lethal punch with collapse about to follow. Some of the contempt went out of his expression as he said, 'Why not look at it like it is, Duke? Here you are, sitting on top of the world, the biggest promoter with the biggest gate since the days of Tex Rickard.' He pointed to the pile of papers on the floor and said, 'Even they had to go for it, didn't they? You bring off a great boost for boxing and feed a lot of kids. Who's going to open his bazoo that he's been stuck—Uncle Nono? Are you kidding? Us? Brother, we're businessmen. And Matilda ain't talking. As for all them big shots that contributed to your Fund and got their names and pictures in the paper, what have they got to complain about? Them and a hundred thousand others saw Dockerty get up off the deck three times and go on to win. They got a run for their money, didn't they? Take the advice of an old-timer, kid, and let it ride.'

'Oh, for Chrissake shut up and get out of here!' Duke Parkhurst said wearily. 'I want to think.'

The three arose and moved for the door. After they had passed through, Patrick Aloysius bobbed his head back inside a moment for his exit line, 'Don't think too hard,' he said.

Parkhurst called, 'Wait a minute, Ahearn! Did you bet on the fight?'

The old manager grinned and said, 'Sure! Everything I had.'

'On whom?' but Parkhurst already knew the answer.

'Who the hell do you think? On Dockerty by a knockout.'

All the contempt which had previously been Ahearn's now gathered on Parkhurst's expression. He was too soul sick even for obscenity. He said, 'You're really just about four stories lower than a louse, aren't you, Patrick Aloysius Ahearn?'

The manager grinned at him, said, 'I wouldn't say that. Put it that I'm a gambler. Don't forget that I came within three seconds of losing,' and he ducked out.

After they had gone, Parkhurst bowed his head in his hands and sat there for a long time. How had he got himself into a position where the cynical, crooked manager of a stable of third-class fighters could mock him, and a little ex-booker of cheap night-club acts could advise him what he could and could not do and, what was more, putting it to him

277

in such a way that he was going to have to comply?

For he knew now that he would never write that story as it was; he would never write it at all, not so much because he would be laughed at himself, but for the sake of the paper for which he worked. It was this paper which, through its charity drives resulting in increased circulation and free publicity, had been willing to become involved in the dirtiest racket in the often sordid business of sport—prize fighting, the public torture of human beings. Anyone who voluntarily entered that cesspool was going to find himself up to his neck in what one usually finds in cesspools.

And what Patrick Aloysius had accused him of with the promotion of his famous amateur tournament had struck home, and hurt as much. The inspiration by which they conducted the tournament was ostensibly the noble one of 'getting the boys off the streets'. But in his heart Parkhurst had always known it was the bunk; it was a circulation and advertising stunt. And even though through all of this his own dealings had been scrupulously honest, the fact was—as both Bimmie and Patrick Aloysius had pointed out—his hands were no longer clean, nor were those of the editor, the manager and the publisher of his paper.

The fact that the charity affair had turned out a rousing success redounding to his and the newspaper's credit did not alter the fact of how it had been achieved. He could not afford to drag the *Daily* and *Sunday Mercury* through the filth that would be thrown if he wrote the truth about the Matilda-Dockerty match. The three had offered him amnesty. They would not talk and he doubted very much whether Uncle Nono, having been played for the sucker of the century, would either.

A further pang stabbed him to the heart: Matilda, that loving, lovable, mystery creature, the greatest boxer in the world as long as he was not tapped; his childish passion for bananas, Hershey Bars and chocolate ice cream; *his* ruthless exploitation at the hands of man, exiled, marooned thousands of miles from his native bush, sold to a murderer...

A wild, ridiculous and utterly mad hunch stabbed at his brain. Uncle Nono, Birdie's 'Joe'. But 'Joe' and 'Gio' were pronounced in exactly the same way. Without thinking

278

further and impelled by rage, frustration and humiliation, Parkhurst searched through the directory, picked up his private telephone and dialled a number. When a feminine voice cooed, 'Neapolitan Import and Export,' he gave his name and asked for Mr Giovanni di Angeletti. A few clicks and he had him on the line.

Parkhurst said, 'Mr di Angeletti? Parkhurst speaking. What are you going to do with Matilda?'

Over the wire came the cultured voice of the importer and exporter, 'Marinate him, Duke. They say kangaroo meat is quite tasty prepared that way. Care to come over for dinner when he's ready?'

Parkhurst's temper went again and he shouted into the mouthpiece, 'You do, you dago son-of-a-bitch, and I'll smear the story of what you fell for all over page one!'

The only reply from the other end was a chuckle and, 'What are you so sore about, Duke? I'm not squawking...' There was a moment of silence followed by, 'You're just the kind of guy who might do that, I suppose. What's your price for keeping the whole thing under your hat?'

The shout was already rising from Parkhurst's guts, 'I haven't got a price!' when something made him choke it off. Was this really the dreaded Capo of the Mafia to whom he was speaking, or was he being kidded? If it was, then perhaps there were more ways than one of skinning a cat. He said, 'Get out of sport and stay out!'

The line went silent again. Then the reasonable voice said, 'That doesn't seem to me to be too high. You've got yourself a deal, Duke.' And the connection was broken.

Justus Clay came charging into his office, all his dourness gone; he was lit up like a Christmas tree. He shouted, 'Duke, boy, you put it over! I've just been talking to the Commander and he's tickled pink.' He noted the telegram on Parkhurst's desk and said, 'You're getting a boost in salary as well.'

For a moment Parkhurst was tempted to make a clean breast of everything to his editor, including the fact that he had established the identity of Uncle Nono. In the face of Clay's enthusiasm he could not. And besides, what did he have to go by to prove who di Angeletti was except a phone call which could be denied?

Clay said, 'You've got it coming to you, Duke. Come and

see me in about an hour on the follow-up stories.'

Parkhurst said, 'Thanks,' and when Clay left, knew that he had thrown in the towel. He wanted to go home and bury his head against Birdie's breast and hide in her arms. The thought struck him that even for her he had Uncle Nono to thank.

The telephone rang in the suite which Bimstein, Ahearn, Baker and Company had leased in the Lincoln Hotel as their headquarters for the big championship. Ahearn picked it up. muttered sotto voce several times into the mouthpiece and said, 'I'll see,' and then, covering it with his hand, said to Bimmie, 'There's a Miss Lebensraum on the phone. She wants to talk to you. Are you in or out?'

Bimmie said, 'Hannah? Oh, my God! She said she never wanted to see me again because I wasn't honest. She gave me the air.'

Patrick Aloysius said, still keeping the mouthpiece well covered, 'A half a million dollars in the bank could change the outlook.'

Bimmie said, 'Hannah ain't like that. She wouldn't care if I didn't have a dime. She used to stake me to meals when I was flat on my tochas. She chucked me because she was worried I wasn't on the level.'

Ahearn grinned wickedly and said, 'Well, she sure had something to worry about. What do I say? You just left for the Coast?'

'No,' said Bimmie, 'gimmy!' and took the phone, cradling it as though it had been Hannah herself. He said, 'Hello, Hannah! It's Bimmie. You calling me?'

Hannah's voice seemed rather small, faint and distant as though the connection were not too good. She said, 'Hello, Bimmie, how are you?'

'Fine! Great! How are you?'

'I'm fine, too,' came Hannah's voice.

Bimmie said, 'That's good.' Then there was a silence.

Finally Hannah asked, 'Are you busy, Bimmie?'

'Well, yes—sort of. We got a meeting kind of, see, after what happened last night...'

'I know,' Hannah said. 'I wanted to talk to you maybe, if you could come over sometime.'

Hope rekindled once again in Bimmie's narrow chest. Was it possible that there could be a reprieve? He said, 'You want to talk to me, Hannah? Okay, I'll be over in twenty minutes.'

It took him fifteen, as he was able to catch an express at the 42nd Street station of the Broadway subway. On the ride he was prey to what were known as mixed feelings, for as he had left the suite he had been followed by Patrick Aloysius' cold stare and those tobacco-stained lips murmuring, 'For half a million dollars, what's honest—eh, Bimmie?'

It was this that was worrying him. If Patrick Aloysius was right then everything else was all wrong, and Hannah Lebensraum was not honest either, because if she thought he was a crook then she should not change because he was a successful crook. He had tried to bamboozle her and give her a line and talk her into accepting him as he was. It was not his own malleable ethics that were worrying him, but through it all Hannah's had something to which to cling. You could come out of the downtown cesspool, take the ride to Upper West End Avenue and there she was, beautiful, clean and fresh like a flower and when you were with her maybe you did not smell so bad yourself any more. And who was worse? He, Bimmie, always trying to justify himself, or Patrick Aloysius who shamelessly and cynically, and with no trouble to his conscience, operated on the lowest level of necessity?

The excising of Hannah from his life had left Bimmie with a feeling of failure and futility and through his mind ran that phrase accompanying the toss of a coin, 'Heads I win; tails you lose.' He had won and he had lost, and the loss was bigger than the gain.

Thus it was full of misgivings such as he had never experienced before that he rang the doorbell of the Lebensraums' apartment. Hannah opened it. She was looking paler than when he had last seen her and there was also an expression of determination. She said merely, 'Come in, Bimmie,' and led him into the same living-room where it seemed only a short while ago she had toppled his world by returning his engagement ring.

They sat down on the couch facing one another and listened for a few uncomfortable moments to the ticking of the ormolu clock on the mantel. There was not a sound from

281

any other part of the house. Whatever was going to develop, Hannah Lebensraum had apparently made certain that there would be no disturbance from parents or sister. She had a difficult task to face and now set about it courageously. She said, 'Bimmie, I saw on TV, and also read what happened last night in the papers this morning. You lost, didn't you?'

Bimmie nodded.

Hannah asked, 'Was Matilda... was he hurt?'

'Not really,' Bimmie replied, 'that was the only really good part. He qui... he got counted out before he really got hurt. He just lost his wind.' And then he added, 'He's retiring.'

Hannah turned her large, liquid eyes full upon Bimmie and said, 'But you're in trouble, aren't you, now?'

Bimmie said, 'I guess so.'

Hannah Lebensraum then spoke the most courageous sentence of her life when she asked simply, 'Bimmie, do you still want me?'

The boy looked at her bewildered. He said, 'Are you kiddin'? Do I still want you? Ain't I always? But what's that got to do with it? You said...'

'Oh, Bimmie,' Hannah said, 'don't make it so hard for me! But when you love someone and they're in trouble, you don't walk out on them.'

A light was beginning to dawn upon Bimmie, a rosy one and he seemed to hear sweet music, the singing of angels as well, but he had to be sure. He said, 'But, Hannah, you said you didn't love...'

'I didn't,' Hannah protested. 'I never said I didn't love you. I just said I couldn't marry you. Oh Bimmie, I've never stopped loving you. I've been so miserable.'

'You mean,' Bimmie said, 'that because I'm in a jam now, you'd marry me even if you thought I was a rotten little crook?'

'You aren't a rotten little crook, Bimmie!' Hannah protested. 'But I know now I wouldn't even care any more. You know when we were happy? When you were broke and I used to buy you dinners and believe in you that someday you could make those dreams you had come true. Now that you're going to be broke again, I'm offering myself, Bimmie, if you still want me. Together we could build something again, maybe. I . . .' She ran down. There was no more that she had to say or could say and she sat facing him, looking a thousand times more desirable than ever she had before, illuminated by courage and love.

The light penetrated to Bimmie and he had a momentary insight to what she was and what she was offering. He had the flash of intuition and good taste not to seize her in his arms or say anything. Instead, he just took one of her hands in his and held it gently, looking at it as though he had never seen it before and stroking one of her fingers with his, while he tried to readjust his thoughts.

And what kept interfering with this was the memory of the figure of Patrick Aloysius Ahearn and the cynical twist of his mouth and Bimmie thought that he would have given anything for Ahearn to have been there so that he, Bimmie, could spit in his eye. She was not cashing in on his wealth; she was offering to share his poverty.

And she was regarding him with those luminous, loving eyes and waiting for something more dramatic and satisfying than just having that finger empty of the engagement solitaire stroked, not knowing that Bimmie was struggling like a fly trapped in a spider's web which the curious twists and turns of the Broadway intrigues had woven about him.

For if he told her the truth, he must lose her again. Not broke; not in trouble; rich and due to be richer; all as the result of probably the biggest triple cross in the history of boxing.

The finger was as soft as velvet and yielding. Hannah's dark hair had a disturbing fragrance that was not of scent but something of its very own. Panic brought out sweat upon his brow. Cornered, trapped, pinned down he fled to something which a few moments before he never would have considered even as a last resort. But there was no other way, not with a girl like Hannah Lebensraum.

He said, 'Hannah, baby, look! I got to say it to you, even if afterwards you turn me out again like you done before. See, what you read in the papers is only one thing about now how Matilda got licked and lost the fight, and won't fight any more. But I . . . We . . . We ain't broke. I ain't really in trouble. I got a half a million bucks and a partnership in Lee

Dockerty, and a business with Patrick Aloysius. We got it made, Hannah, and it wasn't honest—all except for Matilda.'

She did not withdraw her hand. Somehow the gentle stroking of that finger was the most wonderful sensation she had ever experienced, better than kissing or hugging or being in Bimmie's arms. She said, 'Tell me about it, Bimmie.'

He thereupon did, repeating the story he had told Parkhurst, not necessarily about Matilda's build-up but about the last minute deal in which Uncle Nono had bought a dud kangaroo and the three had split the million dollars as well as acquiring Lee Dockerty as a part of the trade.

'So that's how it is, Hannah. I'm a no-good and a fine, honest girl like you shouldn't have nothing to do with a bum like me. Maybe if I didn't love you, I would have told you a lot of lies. So before you turn my hat out the door and ask me I should follow it, maybe I ought to go.'

But Hannah still made no move to withdraw her hand from his. Instead she lowered her eyes and herself began to stroke Bimmie's finger with one of hers. She said introspectively, 'After you went away, Bimmie, I couldn't help thinking about what you said about what's honest. Maybe only when two people say they love one another and mean it, and could stick by it, perhaps that's the only thing in the world that's honest and in which you can trust, and as long as you don't lie and steal or hurt anybody . . .' She played with the knuckle of the finger and added, 'Mr Nono came to you to buy, didn't he?'

'Yes.'

'And you sold him what he wanted to buy?'

Bimmie was the one who was now watching Hannah warily. He said, 'But the goods were phoney.'

Hannah Lebensraum said, '*Caveat emptor.*'

Bimmie said, 'Carve what?'

Hannah translated, 'It's Latin. What it really means is always examine the merchandise before you put down your money. There's no law against it. Let the buyer beware. Maybe the mark-up was a little high, but that's business, isn't it? Bimmie, for the first time in your life you didn't give me schmoos. You told me the truth and I love you for it. I don't care. I know you're a good boy at heart.' She was now looking at him with a wry and enveloping tenderness. 'If you like, Bimmie, you can put the ring back on my finger again.'

Struck aghast, Bimmie dropped the small hand as though it had been a hot potato. He said, 'I . . . I . . . I can't!'

'Bimmie!' Hannah wailed. 'You mean that you don't love me? You don't want me?'

'No!' Bimmie replied in anguish, 'I can't! When you gave it back and turned me out, I thought you meant it forever. I sold it to Valentine's on Fifth Avenue for nine grand. But I could buy it back for twelve.'

At that point Hannah gave way to what she had been wanting to ever since Bimmie had initiated the recital of his iniquities—laughter, to the complete mystification of Mr and Mrs Lebensraum and Myra, eavesdropping down at the end of the hall.

'Oh, Bimmie,' Hannah finally cried when she had laughed herself out, 'you're wonderful! You made a two thousand dollars profit and you want to give them three for me? Do you know what I want from you, Bimmie? Would you do someting for me? I want another engagement ring from you, but I want it from Woolworth's. It shouldn't cost more than ninety-five cents, and I'll wear it with pride for the rest of my life, for teaching me not to be a prig.

And after that there was nothing more between them to keep them from going through those actions of truth and murmurings and the clinging that is necessary to young people when they are in love.

# XXVI

IT WAS TWO weeks later. Parkhurst was at work on his column for the next day in his office. He had been learning to live with his conscience which sometimes even caused him chuckles in spite of himself, but more often qualms, particularly when he wondered or contemplated the fate of Matilda who had vanished from the news and, for all he knew from the face of the earth. If he had truly, as had been threatened, journeyed through Gio di Angeletti's alimentary tract, Parkhurst would never forgive himself. For the rest, time dulled even guilt by default.

The dopey boy from the reception desk shambled in saying, 'There's a guy outside wants to see you, Mr Parkhurst.'

'Who is it?'

'I dunno. He talks funny.'

Parkhurst took the slip of paper and examined it. The scrawl announced his caller as Billy Baker, and in parentheses, the Bermondsey Kid. He said, 'Well, blow me down! I thought he'd headed back for the Thames-side. Send him in.'

Except for a new suit of clothes and hard hat, the Cockney was the same and after shaking hands, gingerly sat on the edge of one of the office chairs.

Parkhurst said, 'Well, hullo! I thought you'd gone back and by now maybe were drawing mild and bitter from behind the bar of the Bunch of Grapes.'

Baker said, 'I'm off next week, sir. I'm just staying on for Bimmie's wedding. I expect you'll have had an invitation yerself.'

'I have,' said Parkhurst, 'genuine engraving. Passes the fingertip test,' and he wondered what was the reason for his visit.

Baker said, 'But before I went, sir, I thought as how you'd like to know about Matilda.'

All of Parkhurst's doubts, fears, forebodings and guilt feelings returned. Had Baker come to reveal some horror that would haunt him for the rest of his life? Still, it had to be faced and he asked, 'Ah yes, Matilda. What about him?'

'He's safe, sir. He's okay. Happy as a sandboy.'

Parkhurst could have kissed the little man, so great a load lifted from his mind. He grinned cheerfully enough now and said, 'Come on, Billy, tell! I kind of got to love that silly animal myself.'

'Well, sir,' said Baker, 'you can imagine, sir, how I felt. Him and me practically dossed down together for near on nine years and never a harsh word between us. My making a good livin', always together through thick and thin, good times and bad. And what do I do? I sells him to a sodding murderer as what tried to kill him once before. It was on me conscience. I hadn't reg'lar closed an eye or slept a wink after that. I had to do something.'

'I can see that. So what did you do?'

'I went and had a word with this Mr Nono.'

Parkhurst was startled enough to drop his quizzical mood and exclaim, 'What? Did you get to see him?'

'That I did, sir.'

'And how did you accomplish that?'

'Through this Mr Renato who made the deal with us. I laid me heart bare. He wasn't a bad chap, he seemed to understand.'

'And he took you to Uncle Nono?'

'That's right, sir.'

'Where was this?'

For the first time a slight caginess came into the eyes of the ex-fighter beneath their beat-up brows and he said, 'I wouldn't know, sir. It was kind of a orfice building.'

'And there you met Uncle Nono?'

'That I did, sir.'

'Did you find out his real name?'

Again the wary look, 'He didn't say, sir.'

'What was he like?'

'A proper toff, sir, a real gent if ever I laid me eyes on one, sittin' there be'ind a swanky desk dressed up posh in a coupl'a 'undred quid's worth of clothes. He gives a nod to Mr Renato who blows himself out and there we are, just the both of us.

'"Mr Baker," he says, and I knows the voice of an educated man when I 'ears it, even though it's American, "Mr Baker, what can I do for you?" So I comes right out with it. "Mr Nono," I says, "have you ... Are you going to shoot my Matilda?"

'"MY Matilda," he corrects me, "I paid for him, you know."'

'I had to give 'im the right of that. "*Your* Matilda, sir." I says.

'"But of course I'm not going to shoot him," he says, "Why should I?"'

'Well, Mr Parkhurst, I can't help thinking of that time down in Clayton, Oklahoma when that there Nickel-Plate and Chimpanzee and the bloke with the fish eyes were going to kill Matilda, if he didn't lay down to that there Cowboy Jones and I just mentioned it like, in passing, to Mr Nono. Do you know what? He finks for a moment and then he smiles at me ingratiatin' like, and says, "Well now, Mr Baker, shall we put that down to a momentary aberration and loss of sense of humor? Of course it was stupid of me to take it out on an innocent dumb animal and I should have instructed them simply to shoot you and those other two chuckle-heads if you couldn't take a hint, and I wish to apologize. As it was, as I remember, the whole affair was badly fouled up and I had to make an example of poor Smiley, who couldn't resist trying to make some easy money. He isn't with us anymore, but his family's being taken care of. We always do."

'Can you imagine, Mr Parkhurst, him comin' right out and tellin' me this fyce to fyce?'

Parkhurst said, 'Well, yes, actually, I can.' He added, 'Had you already delivered Matilda?'

'Yes, I had. I keeps me word, Mr Park'urst, a deal is a deal, even though it broke me heart to say goodbye.'

'How was this done?'

'I took him in a van up to Hawthorne Circle where I'm met by another van with nothing painted on the sides and the licence plates covered up. Four men, not bad lookin' blokes, helped me make the transfer. Matilda, he gives me one last smack: I gave him one last jumbo-sized Hershey Bar; they closed the van and drove off. I was blubbin' like a baby.'

Parkhurst nodded, 'That would be quite the safest and most intelligent way to do it. So what did Uncle Nono have to say to you, after telling you that his intentions toward Matilda in Clayton had been honorable?'

'He says to me, Mr Parkhurst, he says, "And now, Mr Baker, I would take it as a favor if you could let me know exactly the situation with Matilda. I know you've sold me a pup, but not yet what kind of pup. I've been going to fights long enough to know that he wasn't hit hard enough to be hurt. Was he sick? Is he yellow? Can't kangaroos take it alabanza?"'

'And you told him?'

'Just like I told you. The whole bleedin' truth about 'roos and what they're like, that's to say the strange nature of the beast, not excluding the sexual angle and how it is with 'em back home in the bush, and everything that happened from the time he zapped Dockerty that night way back in Camauga.'

'And what did he have to say to that?'

'Nuffink. He just laughed.'

'Laughed?'

'He had a fit worse than you had, sir. I never seed anyone laugh like it before. Slappin' of his sides, wipin' his eyes. He laughed so hard he was making so much noise so some doors opened and a coupl'a tough-lookin' mugs come in to make sure he wasn't being assinsated, or something. They gives me some mean looks but he waves 'em away so he can finish laughin'. But at last he's finished, gives his eyes another wipe with a solid silk handkerchief, leans across his desk and says to me, "Go home, Mr Baker, nothing is going to happen to Matilda. I've got a stud farm; lovely place; some of the best grass in the country. There's a nice warm barn where he can be in the winter, I guess kangaroos don't care for the cold very much. And in the summer, fields where he can graze."

'So then I tells 'im all the rest I knows about 'roos—their habits, their likes, their dislikes, their feedin' and especially about Matilda and his lovin' disposition. "Thank'ee, sir," I says to 'im. "Blimey, but you're a real good cove and what's more, a real gent. Thank'ee again, sir. You've put me mind at ease."'

'And the catch?' Parkhurst enquired blandly.

Baker looked at him startled, 'How would you know there was one, sir?'

'There has to be, doesn't there?' said Parkhurst. 'There always is.'

'There ain't much of one. You wouldn't really call it a catch, sir. Just before I takes me leave an' he shakes me hand, he says, "It's our little secret, ain't it, Mr Baker? That's to say, ours and I suppose those other two, and of course, our mutual friend Mr Parkhurst. As long as you keep your mouths shut and there isn't so much as a whisper of what's happened between us, Matilda's as safe as though he were with you. Talk, and I'll ship him in bits all the way to you in Bermondsey, canned!"—that's American for tinned, sir. Fair gave me the horrors for a moment. But you wouldn't talk, sir, would you? And neither would we. So we shakes hands and 'e's got his finger aimed at a buzzer when he halts it midway, sudden like, and says, "Mr Baker, man is mortal and some might be considered more so than others. If something were to happen to me during Matilda's lifetime, what would you like done with him? Send him to you? Have him shipped back to Australia to his own kind?"'

'And what did you say?' Parkhurst asked.

'"Not Australia, sir," I says to 'im. "It was on me own mind to take him back there and turn him loose in the bush instead of keepin' him in captivity, after we'd made our pile. But that was before the fight. Not now any more. I'd be obliged if you'd send him along to me and he'd never want for nothing. He can't go back to Australia. Don't forget, he's been zapped; them other 'roos would make life miserable for him, him that would 'ave been king of them all."'

'And what did he say to that?' Parkhurst enquired.

'He quoted the Bard of Avon at me, '"Uneasy lies the head that wears a crown". Maybe he's better off as he is. Very well,

I'll leave instructions with me estate to send 'im along to you in case of accident." A proper toff, Uncle Nono.'

'And then?'

'His finger pushes a buzzer. A bird comes in with a shorthand book. He says, "Miss Benson, show Mr Baker out and give him a tin of our special Valdesti Olive Oil and one of those salamis that have just come in from Foggia, as a souvenir. And thank you again, Mr Baker."'

Parkhurst sighed and said, 'And I thank you, too, Mr Baker."'

The Bermondsey Kid got up to go. He said, 'See you at the weddin' next week?'

'I wouldn't be surprised,' said Parkhurst, and alone once more he gave vent to a second long exhalation. He was relieved about Matilda and both irritated and admiring at the manner in which Gio di Angeletti had put paid to any idea he might have had of breaking the story of Matilda. Some day the Forces of the Law were bound to merge Uncle Nono and Gio, and when that happened, he felt he might even be rather sorry. As Baker had put it, if nothing else, the man was a proper toff.


The wedding of Solomon Bimstein and Hannah Lebensraum at Temple Beth Shim, on Broadway and 91st Street was a memorable one, outstanding even in a list of other notable orthodox ceremonies of the year. Taking place at eleven o'clock in the morning it combined the élite of Upper Broadway, Amsterdam and West End Avenues with celebrities and denizens of the midtown Main Stem, many of whom, unaccustomed to surfacing at that hour, had practically to prop up their eyelids with matchsticks to keep awake during the service.

When, however, the young couple had been joined indissolubly and the guests descended into the basement of the Temple, they woke up to consume caviar, lox, sturgeon, blinies, bagels, carp, herring and other delicacies and to learn to dance the hora to the strains of Meyer Davis's orchestra. For, seeing that the groom was on his way to becoming a millionaire the affair was as lavish in food, décor and souvenirs for all as any within memory of the oldest inhabitant of the neighborhood.

Outside the portals of the Temple stood a gleaming white Cadillac roadster with red leather upholstery and dashing chrome—Bimmie's wedding present to his bride. Behind it was a more modest sedan—his gift to the bride's parents. His own family he had already taken care of with a more luxurious apartment, as befitted half of the managerial partnership of Ahearn and Bimstein. For as success produces success, other famous fighters were now clamoring to become members of that stable and someday they might even find themselves representing the heavyweight champion of the world.

Mr and Mrs Duke Parkhurst, Murray O'Farrell and Justus Clay of the *Mercury*, and most of the sports editors and boxing writers from the other newspapers came. Patrick Aloysius Ahearn and Billy Baker, of course, were there along with Lee Dockerty. Pinky Schwab was not present; he was said to be suffering from a nervous breakdown brought on by the news of the partition of Joe Marcanti. But Gentleman Johnny Donohue of Donohue's Gym and most of his crew of scrambled ears arrived, along with boxing buffs, Walter Mason, President of the Arenas and some of his society patrons, actors and actresses from stage and screen. Warmhearted, generous, unjealous, Broadway rejoiced in welcoming a new rising star—Bimmie.

The families of the happy couple made immediate friends. The Bimsteins expounded upon what a beautiful and accomplished girl was Hannah; the Lebensraums reciprocated with what a fine and clever boy Sol had turned out to be.

It even included an actual scene from one of Bimmie's fantasies when he was poor and a struggling unknown, when Mr Lebensraum proffered an open box of Havana cigars to a famous film director and said, 'Have a Hoyo di Monterey. You can't get them here any more, but my son-in-law had them sent over from England. By air, yet.' It was a wonderful ending to a beautiful and unusual success story which showed that in spite of everything, if a young man today in New York had it in him, he could become rich and famous.

Duke Parkhurst, towering over Patrick Aloysius and The Bermondsey Kid, stood on one side of the room with Birdie, contemplating the dancing couples who had now lapsed from the ancient Hebrew folk dances to the newly returned, modern, dreamy, cuddling-close numbers that youth had only recently discovered.

Patrick Aloysius said, his voice always tinged with the faintly sardonic, 'Terpsichore has no charms for you? You don't trip the light fantastic?'

Parkhurst said, 'If you had feet the size of mine, you wouldn't risk depositing them on something as fragile as my wife. Besides which, she was a professional dancer and I'm scared.'

Patrick Aloysius said, 'Yes, I read about your announcement. Congratulations. But why the hell didn't you tell me before, in Chicago? You nearly gave me heart failure.'

Parkhurst merely grinned at him. He said to Billy Baker standing nearby, 'You all right, Billy?'

Baker nodded and said, 'It's kind of lonesome, like, without Matilda but as long as he's happy, I'm happy.'

Parkhurst asked, 'How do you know he's happy?'

The little ex-boxer looked about to see that no one was too nigh, then stood on tiptoes and whispered to him, 'I've had a secret visit to 'im. Livin' like a bleedin' king, that's what he is.'

For a moment the newspaperman in Parkhurst sparked, even though he knew he would never make use of it. 'Where was it?' he asked. 'Could you find it again?'

'Not a chance!' replied Baker. 'Proper cloak-and-dagger stuff, like in the flicks. Closed car, me in the back with a bandage around me eyes, ridin' for a couple of hours and no tellin' where I was goin'. But when I gets there it's a paradise for man and beast, I'm tellin' you.'

'And Matilda?'

Billy Baker showed his broken-toothed grin. He said, 'You know what the sod did? He went and hid 'imself be'ind some bushes when he saw me. He was afraid I was comin' to take 'im away.'

Parkhurst exclaimed, 'Why, the ungrateful...'

But Billy Baker shook his head. 'How could you blame him? It was me as sold him, wasn't it? And besides, he'd never had it so good. Do you know what that there Mr. Nono done?'

'No,' said Parkhurst.

'Offered to sell him back to me!'

'For how much?'

'One dollar and other considerations.'

'What were the other considerations?'

'"It's just a legal term," he says, but I should also keep me ruddy trap shut. I says, "No, sir, thank you, sir. But he'll be happy 'ere with you, out in the open. We'll just stand by our original agreement."'

'When Matilda seed me back in the limousine to go home, he poked 'is 'ead around the bushes. They put the bandages back around me eyes and that's the last look I had of him.'

And the motivator of it all? Matilda, the big gray, doe-eyed kangaroo, an expression of ineffable contentment upon his long, furry face was grazing on the fresh, cool grass of a field in Upper Westchester County, against a background of green and white, cupola-topped barns and stables with further away, a veritable palace of Colonial pillared and porticoed mansion.

On a white-painted, six-bar fence that surrounded the field sat a distinguished-looking, middle-aged gentleman in dude cowboy clothes; handmade, finely-worked, high-heeled boots, worn-in jeans, purple silk shirt with embroidered initials 'G di A' in yellow and yellow silk neckerchief and, out of that sense of humor that he did not need to suppress on his own premises, topped off by a black cowboy hat on the back of his head. He watched the animal at its grazing.

He climbed down from his perch and strolled over to Matilda who, at his approach, reared up on his strong hind-legs and tail, but with his two stubby forepaws hanging loosely at his chest. Giovanni di Angeletti struck a boxer's stance, his left extended, his right thumb brushing his nose and pretended to jab at the kangaroo. Matilda did not assume any defensive posture but only took a backward hop with an expression of combined alarm and bewilderment.

Di Angeletti dropped his hands, chuckled and said, 'Okay, okay! Relax! I'm not going to hurt you.' He went over to him slowly, so as not to frighten him further and ruffled the fur on the top of his head. Matilda liked that, placed his forepaws lovingly about the man's shoulders and gave him a kiss, until di Angeletti untangled himself from the embrace and backed away, still chuckling as he contemplated his newest possession. He said, 'Maybe you're smarter than any of us, you silly bastard. You know when to quit.' He resumed his seat on the fence top. Matilda had returned to his grazing but was distracted by the fact that for the third time in a brief space he found himself being apostrophized by one of those inexplicable, other kind of kangaroos. For Gio di Angeletti was inclined to talk.

He said, 'Aren't you ashamed of yourself? You got licked, you know. The first time the going got rough you turned rabbit.'

The animal took two or three awkward little hops forward on all fours, bared its teeth and pulled at a particularly luscious tuft of grass. Was it possible, di Angeletti asked himself, that this ungainly-looking beast had actually come within three seconds of winning the undisputed middleweight championship of the world? No, it was too ridiculous, all rather like one of those scrambled dreams for which you go to a head-shrinker at fifty dollars an hour to unscramble. And yet what was more dreamlike than that Giovanni di Angeletti should also be Uncle Nono, and like any of those poor slobs he sent to extinction, might be tipped the black spot himself any time Il Supremo of Cosa Nostra decided that he had outlived his usefulness?

And yet the two tangibles to prove that it was not a dream were there—Matilda and himself.

Matilda left off his munching to stare, half erect, at the man on the fence.

'You can look,' said di Angeletti, 'you cost me a million dollars. And do you know something? A million dollars isn't too much to pay for the best laugh I've had since I left college. It's been a long, long time. You're supposed to keep a straight face when attending funerals. When I was a young man, I used to laugh all the time and I could make others laugh too, the loudest when the joke was on me. Maybe a million bucks is cheap enough for the reminder.'

The kangaroo now sat up straight, cocking both ears forward and di Angeletti was remembering the time of Matilda's potency, when he had the middleweight champion of the world on the floor and within seconds of losing his crown. Aloud he said, 'You were almost a king once, weren't you? Would you have been any happier?' And then he added, 'But nothing lasts, does it? I suppose I should have given you back to Billy Baker, but I rather think I'd like to keep you around as a remembrance.'

He paused and through his head flashed certain signs and signals he had noted during the past few months. If he had read them rightly, things were closing in on him not only from inside but from outside the organization as well, as he knew some day they must. And this last caper, if it reached certain ears, would not be helping matters.

He climbed down off the fence, went over and patted Matilda's gray rump and said, 'You may be seeing Billy Baker sooner than you think, old fellow.'

Then, shoving the black, ten-gallon hat further onto the back of his head, he walked thoughtfully off in the direction of the palatial mansion.

Later, at the wedding, Patrick Aloysius said to Parkhurst, 'It's a great day for our boys, isn't it?'

'Our boys?'

'Yeah, Bimmie and Lee. You wait and see Dockerty under changed management. Hidden charms will suddenly be revealed. And he'll goddam well fight when and where I say he does.' His attention was suddenly drawn to the other side of the room. He said, 'Now there's a groovy looking chick. I'll just go over and give her a treat by introducing myself and inviting her to join in this fandango.'

He left, but Parkhurst had got the message. Bimmie and Lee Dockerty had become, 'our boys'. In other words, his and Parkhurst's boys too, and there was no mistaking what had been on Patrick Aloysius' mind; the past was the past and he expected the columnist to drop his feud with Dockerty.

Birdie's hand stole into his and she said, 'Isn't he a funny man? But look, he dances beautifully!'

Parkhurst clung to the tiny fingers and thought what was the use of persecuting Dockerty further? If nothing else, he had proved his courage on the night of the 4th July and he took refuge behind the small comforting thought that perhaps out of this experience he might have acquired the one ingredient that had been missing in the thunderous daily indictments of his writings, namely some tolerance, the substitution of an occasional gray for his uncompromising blacks and whites.

And that morning there had been a straw in the wind, an A.P. despatch from Memphis, Tennessee, to the effect that the Memphis Martials, the pro football team, was on the block and would be sold. This was the suspiciously up-and-down squad rumored to be a part of Uncle Nono's holdings. Was Gio di Angeletti keeping his end of the bargain? If the Mafia influence was truly to be removed from professional sport, Parkhurst could write 'mission accomplished' across his conscience, even if no one ever knew how it had been done, and wait for the cops to do the rest.

Dancing, enfolded in Bimmie's arms, her head, bridal-wreath crowned, resting on the shoulder of his expensive tailor made suit, Hannah, too, wondered whether it might not all be a dream: the roadster outside, the ring on her finger to which had been added a ruby and diamond bracelet and diamond clip, all the famous and beautiful people as guests at her wedding, the gifts to her parents, the marvellous generosity and the terrible expense of it all.

One small, faint cry still arose from her conscience. For an instant she removed her head from Bimmie's shoulder, leaned back slightly and looked into his eyes. 'Oh, Bimmie,' she whispered, 'from now on, you will be honest, won't you? Always?'

It was like stimulating one of Pavlov's laboratory dogs and the reaction was immediate. 'Baby,' began Bimmie, 'today, what's honest? See . . .'

But Hannah was not going to go through all that again. Accept the dream, accept this man who was being so kind to her. Stop asking questions. Bimmie would be as honest as he could, as honest as anyone could be in this day and age. Enjoy the fire of the gems sparkling upon your wrist and hand resting on Bimmie's arm, put your head back where it belongs, lose yourself in the wonderful music of Meyer Davis.

'Never mind, Bimmie,' she said, as she prepared to do so, 'I know you're a good boy. You're wonderful!'

# BERKLEY BESTSELLERS YOU WON'T WANT TO MISS!

DRY HUSTLE            (03661-8—$1.95)
  by Sarah Kernochan

THE FIRST DEADLY SIN      (03904-8—$2.50)
  by Lawrence Sanders

THE GHOST OF FLIGHT 401   (03553-0—$2.25)
  by John G. Fuller

THE ONCE AND FUTURE KING  (03796-7—$2.75)
  by T.H. White

THE POSSIBLE DREAM       (03841-6—$2.25)
  by Charles Paul Conn

A SEASON IN THE SUN       (03763-0—$1.95)
  by Roger Kahn

THE TETRAMACHUS        (03516-6—$1.95)
COLLECTION
  by Philippe Van Rjndt

TOTAL JOY              (03800-9—$1.95)
  by Marabel Morgan

---

Send for a list of all our books in print.

---

These books are available at your local bookstore, or send price indicated plus 30¢ for postage and handling. If more than four books are ordered, only $1.00 is necessary for postage. Allow three weeks for delivery. Send orders to:

      Berkley Book Mailing Service
      P.O. Box 690
      Rockville Centre, New York 11570